管派古琴制作与修复

Making and Restoration of Guan-School Guqin

下册

Volume 2

昭闻 著

by Zhao Wen

Convention for the Safeguarding of the Intangible Cultural Heritage

United Nations Educational, Scientific and Cultural Organization

Intangible Cultural Heritage

The Intergovernmental Committee for the Safeguarding of the Intangible Cultural Heritage has incorporated

The Guqin and its music

in the Representative List of the Intangible Cultural Heritage of Humanity upon the proposal of China

Incorporation in this List contributes to ensuring better visibility of the intangible cultural heritage and awareness of its significance, and to encouraging dialogue which respects cultural diversity

Date

4 November 2008

Director-General of UNESCO

陕西省哲学社会科学重大理论与现实问题研究项目

Research Project on Major Theoretical and Practical Issues in Philosophy and Social Sciences in Shaanxi Province

程刚省级技能大师工作室学术成果

Academic Achievements of Cheng Gang Skill Master Studio in Shaanxi Province

中国书店

出版委员会

Editorial Board

昭闻　著

by ZhaoWen

目　录
Contents

第一章 分析会诊

Chapter I Analysis and Consultation

琴面底大部分开胶脱落；底板有3至5毫米宽通体裂缝，项、尾、天地柱位置开孔不合理，腰、尾等处破损；除岳山以外，其余配件缺失；整体无灰胎，仅薄层面漆。

Most of the glue on both the surface and bottom plates of the qin falls off; there is a 3-5mm crack in the whole body of the bottom plate, unreasonable open pores in the positions of the neck, tail,and heaven and earth pillars, and damages in the waist and tail parts; except the bridge, other accessories are missing; the whole body has no lacquer cement left, with only a thin layer of paint.

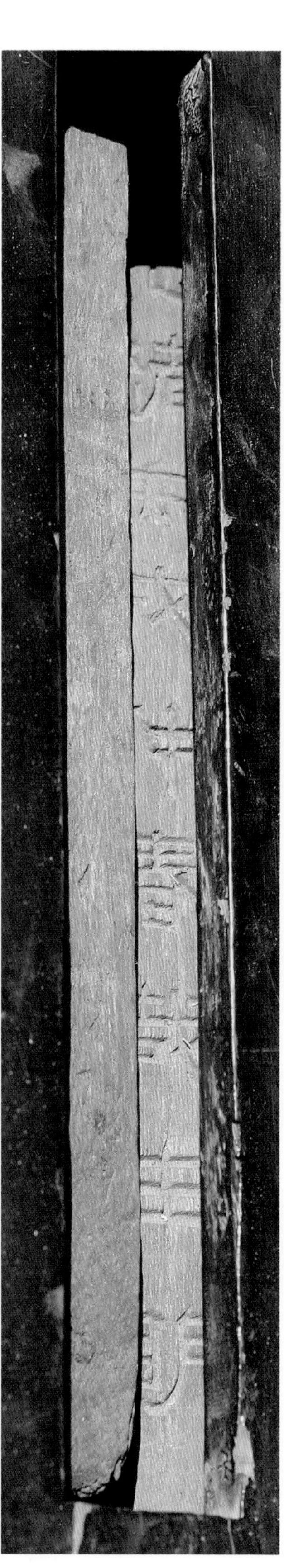

龙池内两侧刻款。

Stamps on both sides of Longchi.

琴额木质腐朽严重。

The wood of the qin forehead is seriously decayed.

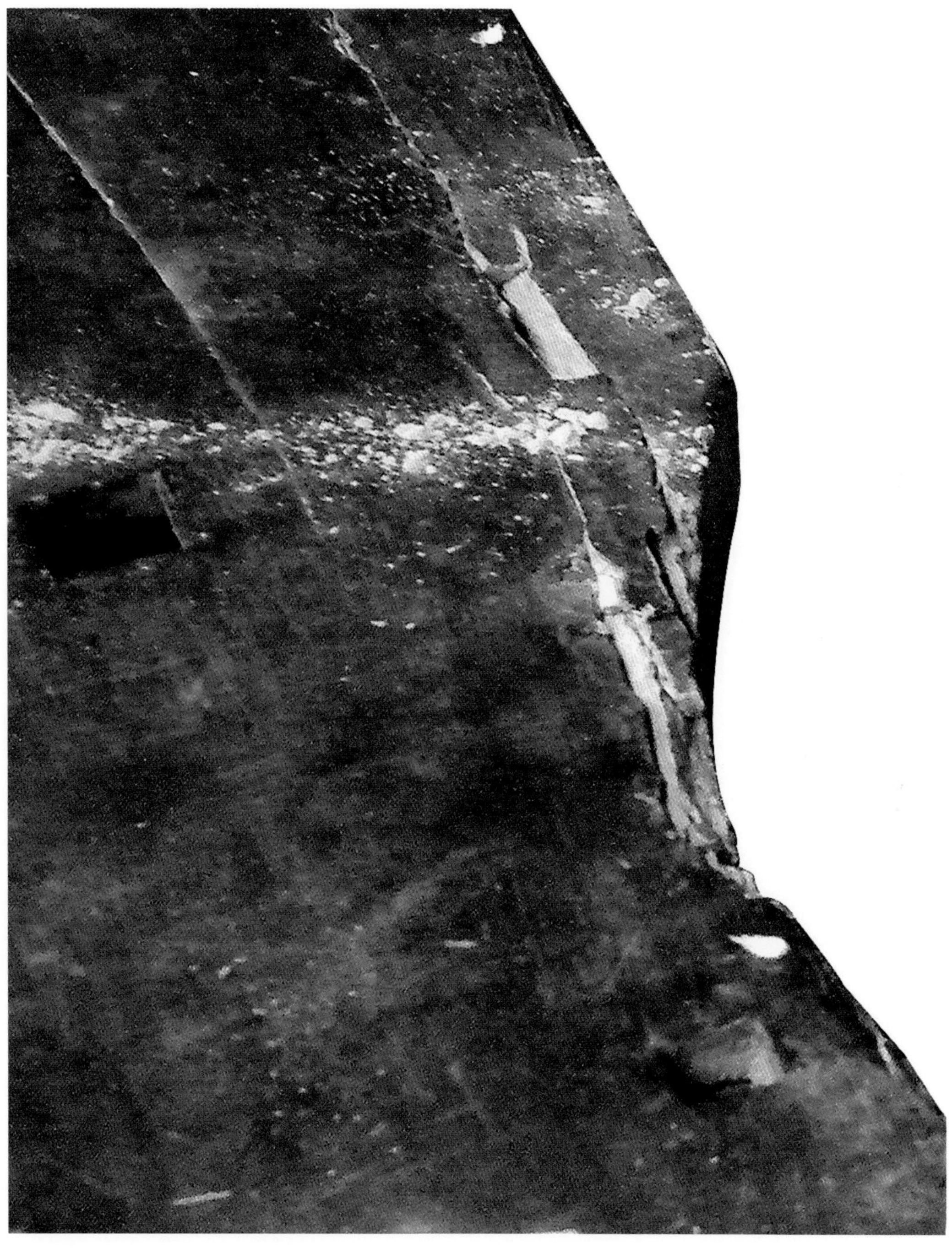

底板腰部破损。

The waist of the bottom plate is damaged.

老琴修复讲究修旧如旧、声音如旧，外观要根据情况而定，有断纹和老漆面的应尽量保留。实在损坏严重影响弹奏的再进行髹漆处理，琴面弦路以外能不髹新漆的则不髹新漆。

一张待修复的老琴拿到手以后，需要整体观察分析，找出问题所在。比较复杂的情况可以邀请同行业专家、材料学专家、文物修复专家共同研究会诊，制定科学合理的修复方案，切不可盲目动手修复，以免导致不可挽回的损失。

本次需要修复的这张清代仲尼式老琴无灰胎，仅有薄层面漆，局部脱落，琴头、尾木质腐朽比较严重，面、底黏合部位大面积开胶脱落，底板 3 至 5 毫米宽通体开裂，腰、尾等处破损，除岳山以外护轸、冠角、龙龈、龈托等配件缺失。从底板项、尾、天地柱位置出现的不合理的臆造开孔来看，此琴应属于所谓的“野斫”，但因时间久远，琴音上好，值得修复。

此琴虽为清代所斫，但并无断纹，其原因是无灰胎，且漆面过薄，无法形成断纹。过薄漆面因长期风化、外力等因素出现脱落，特别是较软木质的琴面，长期磨损更容易脱落。失去漆面保护的木质也就更容易腐朽。

腐朽严重部位以相同或者相近木材进行替换，但能保留原材质的应尽量保留，采用稀释的生漆多遍渗透的方式，增加已腐朽木材的强度，然后施以灰胎、面漆进行保护性修复。

开裂、缺损、孔洞采用相应的方法进行修复，并施以灰胎、面漆，在修复中注意保护原有老漆面，以便完工以后无须修复的部位保持原貌。

When repairing an aged qin, attention should be paid not to change its timber and sound. But the appearance depends on the situation. Those with broken patterns and old paint should be kept as much as possible. If the damages were too serious that affects playing, then the qin should be painted again. Except for the qin surface, other parts may remain unpainted if not necessary.

After getting an aged qin to be repaired, we need to observe and analyze it as a whole to figure out its problems. For more complicated situations, we can invite experts of the Guqin industry, materials science and cultural relics restoration for consultation sessions to formulate scientific and reasonable restoration plans. Never start restoration blindly, otherwise, irreparable damages may be caused.

The Qing Dynasty Zhong Ni-style old qin to be repaired this time has no lacquer cement, with only a thin layer of paint, which falls off locally. The wood at the head and tail parts of the qin decay seriously. The bonding parts of the surface and bottom come unglued. The whole body of the bottom plate is cracked 3-5mm wide, and the waist and tail parts are damaged. Except for the bridge, accessories such as huzhen, crown angle, dragon's gums and gums tray are missing. Judging from the unreasonable open pores in the positions of the neck, tail and heaven and earth pillars of the bottom plate, the qin should belong to the so-called "wild qin". Despite the long ages, it has a good sound, hence is worth repairing.

Though made in Qing Dynasty, this qin has no broken patterns, that's because it has no lacquer cement and the paint surface is too thin to form such patterns. The thin paint surface falls off due to long-term weathering, external forces and other factors, especially at the softer surface plate, which is easier to fall off due to long-term wear. And the wood without paint protection is more likely to decay.

Seriously decayed parts shall be replaced with wood of the same or similar type, while retaining the original material as much as possible. The strength of decayed wood can be increased by diluted raw lacquer, and then lacquer cement and surface paint can be applied for protective repair.

Cracks, defects and holes can also be repaired in this method, and then applied with lacquer cement and surface paint. Attention shall be paid to protect the original old paint surface during the repair so that the parts that do not need to be repaired shall remain the original appearance after completion.

第二章　裂纹修复法

Chapter II Crack Repair Method

在凤沼内临时垫入楔子，将塌陷的底板抬起，然后使用竹钉固定、胶水黏合。

Temporarily put a wedge in the phoenix marsh to lift the collapsed bottom plate, then fix it with bamboo nails and glue it.

将面底板开胶部位上胶，然后再打入竹钉，竹钉呈斜角打入底板开裂部分，起固定作用。

Apply glue on the degumming parts on the bottom plate, and then drive bamboo nails to the cracked part at an oblique angle for fixation.

将用竹钉固定好的裂缝内填入灰胎，填入时力量需要适中，过重会溢入槽腹内，过轻则达不到裂缝内部。

Fill the crack fixed with bamboo nails with lacquer cement with moderate force. If it is too heavy, the lacquer cement may overflow into the groove abdomen, but if it is too light, the lacquer cement may not reach the inside of the crack.

裂缝较宽的地方可预先填充木条或者竹片。

Wood strips or bamboo chips can be filled in advance in wide cracks.

等待灰胎干燥。

Wait for the lacquer cement to dry.

琴体开胶、脱落、开裂是古琴最常见的问题，特别是在北方，空气干燥、早晚温差大，室内暖气和空调都是导致古琴开裂的主要原因。唐、宋时期的琴多裱有苎麻制作的葛布，可以有效防止开裂、变形等问题。

这张仲尼式老琴无灰胎、无裱布，面、底板黏合处开胶、脱落的时间较长，而且木质有腐朽情况，缝隙内聚集尘土较多，上胶黏合前需用棉签蘸 75% 酒精对缝隙内的尘土进行清理，棉签清理不到的地方可以使用吹灰器进行清理。酒精挥发以后再使用酒精、生漆 3 ∶ 1 混合液，用棉签进行涂抹，棉签不能涂抹到的地方可使用针管往内滴入。干燥以后根据木质腐朽程度可多次重复此步骤，直到腐朽木材固化，不掉粉末，并有一定强度为止。上胶重新固定、黏合，然后在面、底板之间钻孔，打入沾胶的竹钉。

底板缝隙也需进行清理，然后呈斜角钻孔，打入沾胶的竹钉进行固定，凤沼部位的底板塌陷部分使用楔子垫平。闭合的缝隙用胶黏合，较宽裂缝使用相同材质木条和灰胎进行修补。修补时将灰胎填入的力量一定要适中，不可过重也不可过轻，以灰胎刚好填入缝隙内部为宜。

The most common problems of Guqin are degumming, falling off and cracking due to dry air, the large temperature difference between morning and evening, indoor heating and air conditioning, especially in the north. Most of the qin of Tang and Song Dynasties are framed with ko-hemp cloth made of ramie, which can effectively prevent cracking, deformation and other problems.

This Zhong Ni-style old qin has no lacquer cement and no mounting cloth. The adhesive parts at the surface and bottom plates are glued and fall off for a long time. Moreover, the wood is decayed, with dirt accumulating in the cracks. Before gluing and bonding, clean up the dirt in the cracks with cotton swabs sticking 75% alcohol. Use an ash gun to clean the parts the cotton swabs cannot clean up. After the alcohol volatilizes, apply the 3: 1 mixture of alcohol and raw lacquer with cotton swabs for the second time. Where the cotton swabs cannot be applied, a needle tube can be used to drip the mixture in. This process can be carried out many times according to the decay degree of wood until the decayed wood solidifies, does not fall powder, and has a certain strength. Apply glue to re-fix and bond the surface and the bottom plates, then drill holes between them, and drive bamboo nails stained with glue.

The cracks in the bottom plate can also be cleaned, then drilled holes at an oblique angle, and bamboo nails stained with glue are driven to fix it. The collapsed part of the phoenix marsh in the bottom plate is flattened with wedges. Glue closed cracks and repair wider ones with wood strips of the same material and lacquer cement. Fill the lacquer cement gently, the force should be not too heavy or too light, to the extent that the lacquer cement is perfectly filled in the cracks.

第三章　缺损修复法

Chapter III Defect Repair Method

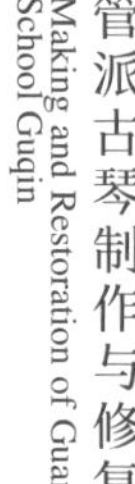

用刻刀剔除琴尾残损部位的腐朽部分，寻找与琴材相近的材料进行黏合修补。

Use a carving knife to remove the decayed and damaged parts of the qin tail, and replace them with similar materials for repairing.

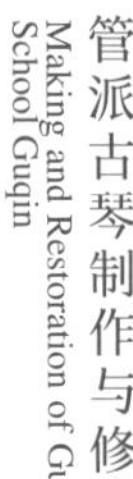

待胶干燥黏合紧实以后修整到位。

Trim such parts after the glue is dry.

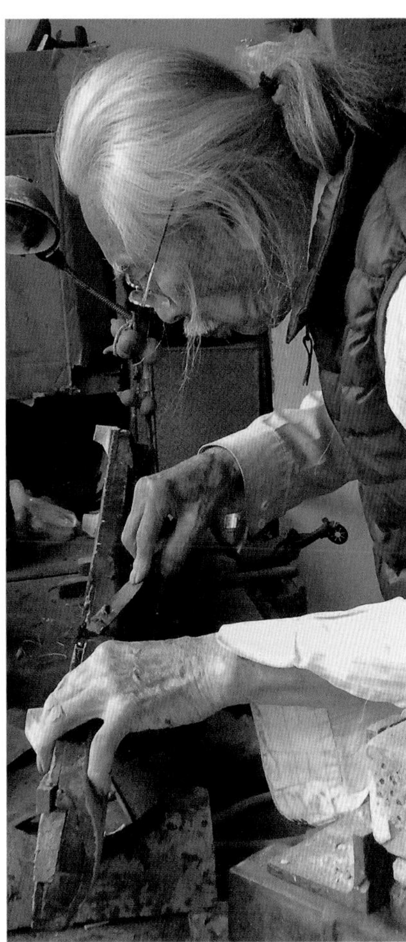

腰部缺损较小地方可用灰胎进行填补。

For small defects in the waist, fill them with lacquer cement.

老琴的缺损多是腐朽、虫蛀、外力破坏等原因所致，腐朽、虫蛀的老琴在修复中稍有不慎，极易出现二次损坏。工作中一定要小心谨慎，避免出现不可挽回的损失。

腐朽、虫蛀的琴非受力部位如果结构尚存，采用酒精稀释生漆进行多遍渗透的方式增加腐朽、虫蛀木材部位的强度。前几遍生漆、酒精比例控制在1 ∶ 3左右，有利于向木材内部的渗透，往后生漆浓度逐渐增加，最后用生漆封闭表面进行保护，稍大一点的虫蛀孔可用注射器向孔洞内注入生漆进行加固。腐朽、虫蛀严重和受力部位无法直接修复的，使用相同或相近材料进行更换，更换前先清除不可修复的腐木，对结构尚存的部分采用前一步骤方式，用稀释的生漆进行加固处理，然后再安装新更换的材料，粘接、打磨、做漆……

一般外力破坏导致的缺损，有的部位很小，直接用灰胎修补即可。有的出现大面积缺失，特别是一些搁置很久的待修复老琴，因时间原因缺失部分会有很多，这就需要综合分析，该粘接的使用胶、漆灰等进行粘接。该修补的，寻找相同或相近木材进行修补，然后按照原琴灰胎的成分、厚薄，制作新灰胎。新灰胎制作时一定要保护好原有的老灰胎，不得覆盖、污染、磨损等。

The defects of aged qin are mostly caused by decay, moth-eaten, external forces, etc. The decayed and moth-eaten qin is prone to secondary damages if not repaired carefully. Be careful not to cause irreparable damages.

For decayed and insect-eaten qins, if the structure of non-stressed parts is well-preserved, apply raw lacquer diluted by alcohol for multiple times on the qin body to strengthen the decayed and insect-eaten wood. For the first few times, the proportion of raw lacquer and the alcohol should be kept at about 1: 2, which is conducive for the mixture to penetrate into the wood. Afterwards, the concentration of raw lacquer should gradually increase. Finally, apply a layer of 100% raw lacquer on the surface for the sake of protection. The slightly larger worm-eaten holes can be reinforced by injecting raw lacquer with a syringe. If the decayed, seriously insect-rotten and stressed parts cannot be directly repaired, replace them with the same or similar materials. Before replacement, remove the irreparable rotten wood first. Reinforce the remaining parts of the structure with the method mentioned in the previous step with diluted raw lacquer. Then install, bond, polish and paint the new materials...

Generally, defects caused by external forces are small in size and can be repaired directly with lacquer cement. Some aged qins to be repaired, especially those that have been shelved for a long time, have large missing parts due to certain factors, should be analyzed comprehensively first, if bonding is needed, then bind by gluing or applying lacquer cement. In case that a repair is needed, look for the same or similar wood for repairing, and then make new lacquer cement according to the composition and thickness of the original. When making the new lacquer cement, make sure to protect the original lacquer cement and not cover or pollute it.

第四章 孔洞填补法

Chapter IV Hole Filling Method

用木工锉打磨因开裂变形等原因导致的孔洞不整齐部位。

Grind the irregular parts of holes caused by cracking and deformation with a carpenter file.

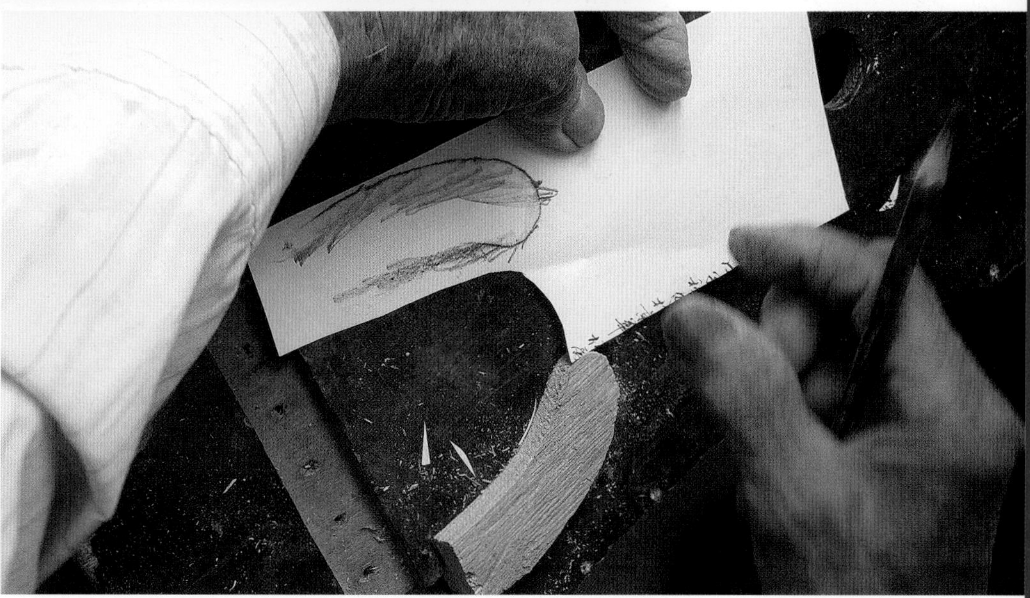

用白纸和铅笔描出需要修补的孔洞大小形状的纸样。

Draw the pattern of the hole to be filled with a pencil on white paper, presenting the size and shape.

用剪裁下来的纸样在提前配对好的软硬相同的填补材料上画出轮廓。

Cut the hole pattern to make the filling material with proper hardness and softness in advance.

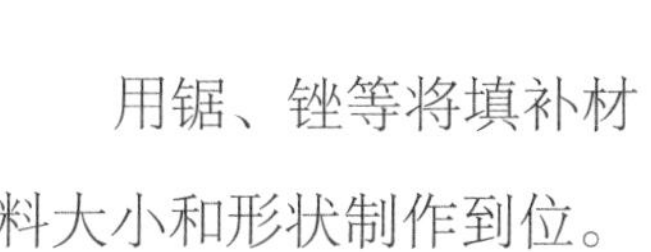

用锯、锉等将填补材料大小和形状制作到位。

Use saws, files, etc. to polish the filling material.

将打磨到位的材料填入孔洞进行测试，用铅笔标出需要进一步打磨的地方和需要裁切掉的过厚部分。

Fill the polished material into the hole for testing, and mark the places that need further polishing and the over-thick parts that need to be cut off with a pencil.

用锯裁切除过厚部分。

Cut the over-thick part with a saw.

将制作到位的材料用胶进行粘接，粘接部位稍有不够严密的，可以使用胶水混合锯末进行填充。

Bond the material made to the hole with glue. If the bonding part is not tight enough, fill in some sawdust or glue.

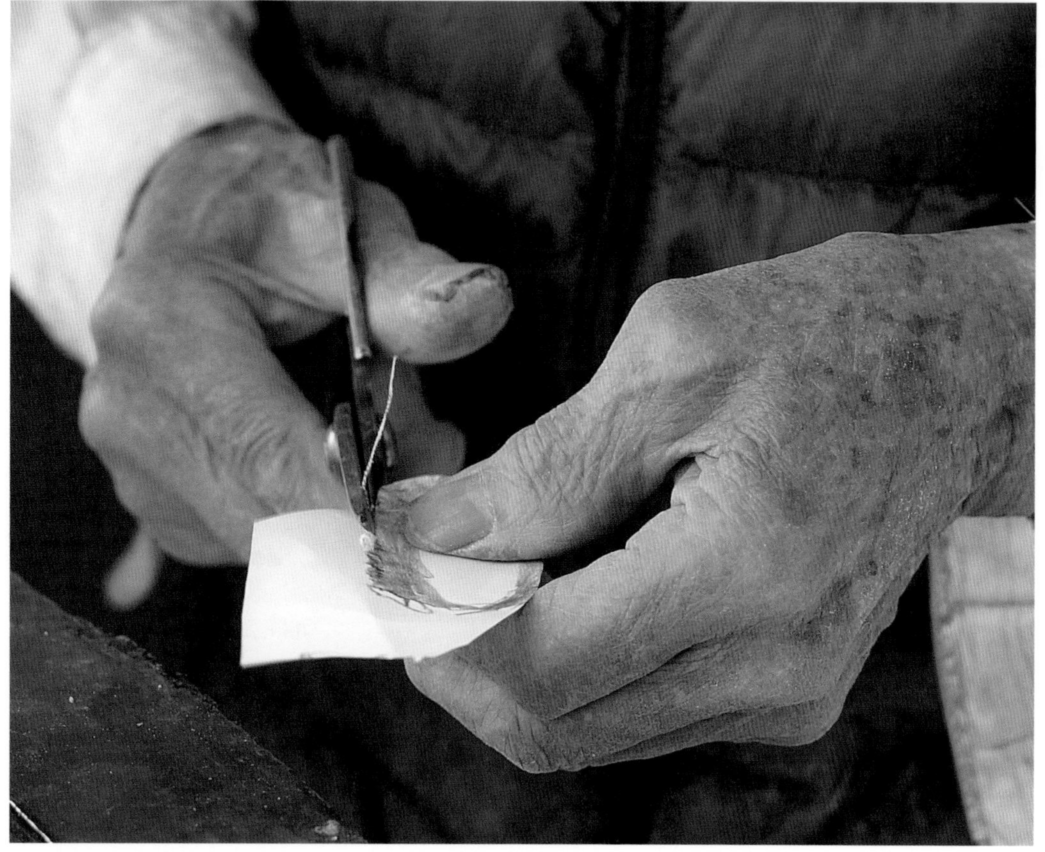

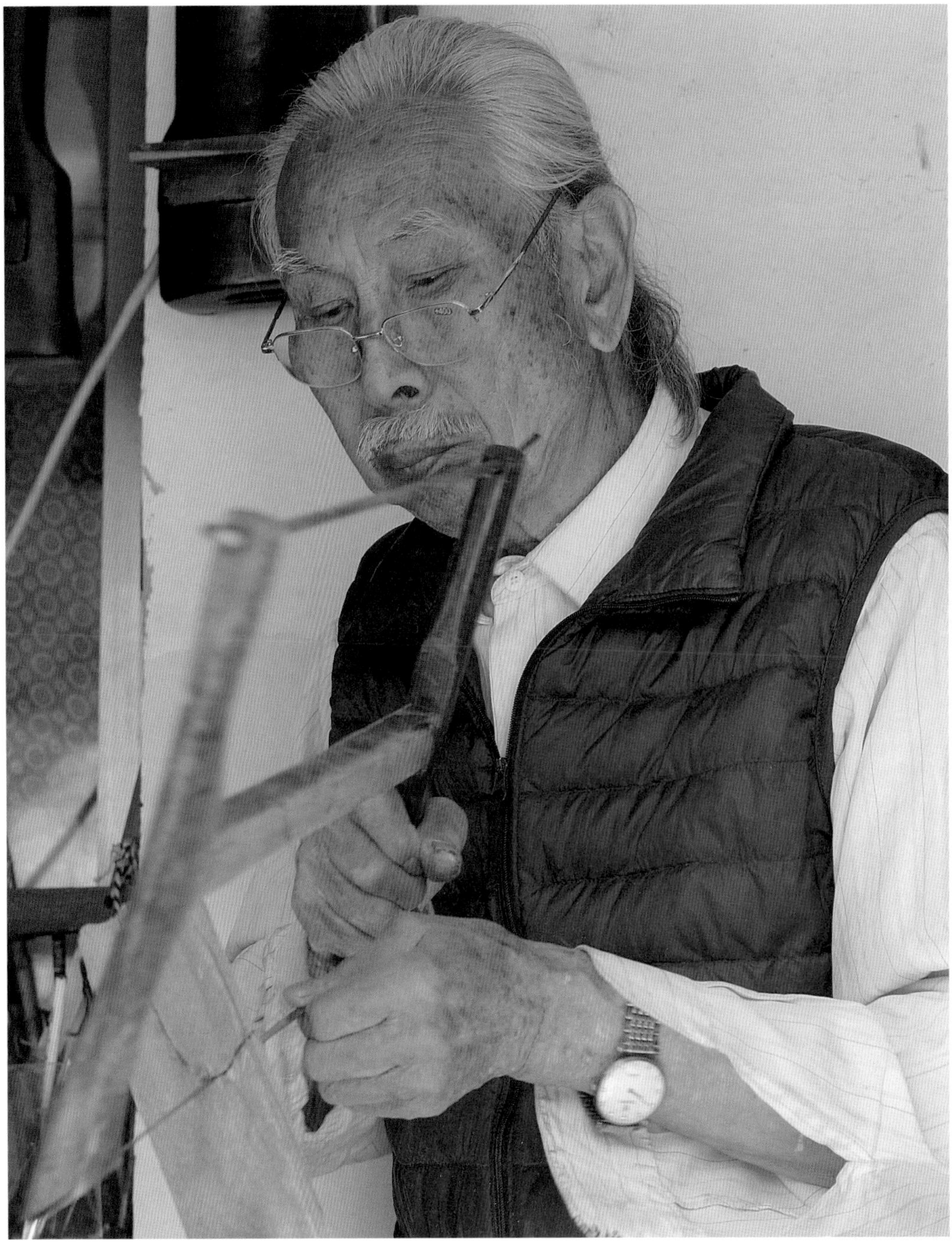

使用同样方法制作琴尾较大孔洞位置的填补材料。

Make the filling material for the larger holes at the qin tail using the same method.

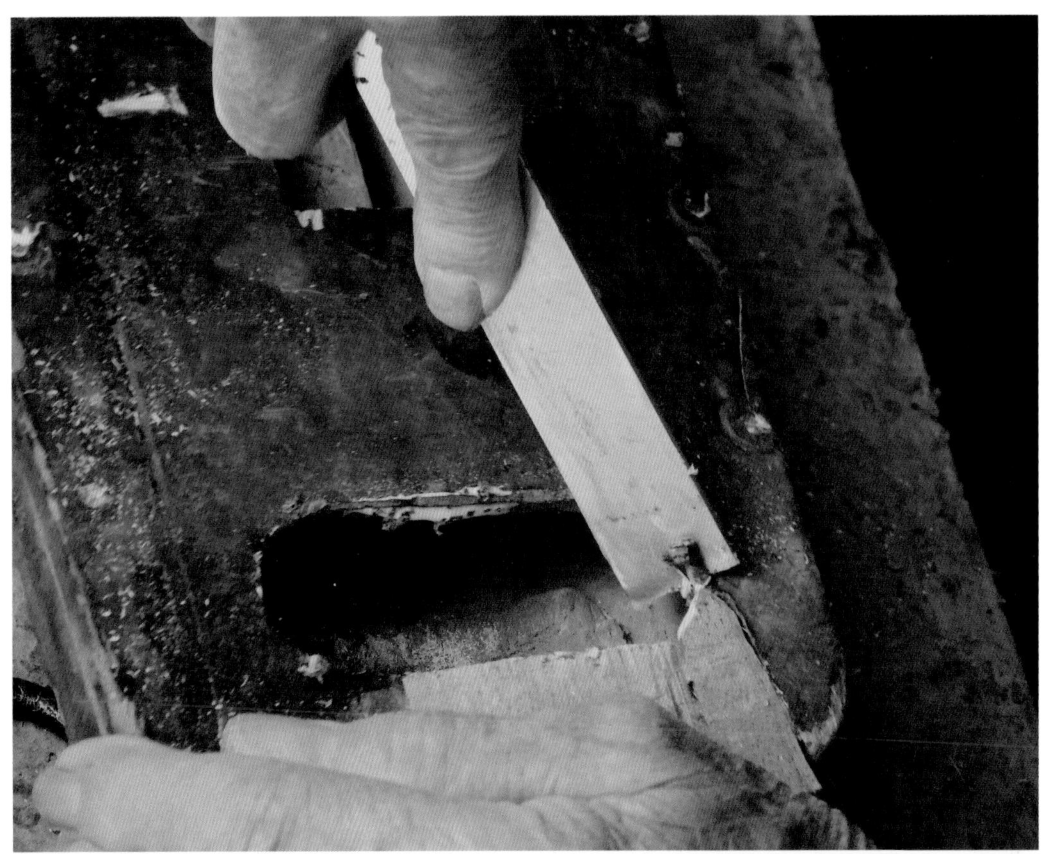

琴尾孔洞较大，而且处于龈托受力位置附近，故将孔洞打磨成喇叭口状，填补材料制作成内小外大，防止在外力作用下出现内陷的情况。

As the hole at the qin tail is large in size and located near the force position of the gingival support, so the hole is polished into a bell mouth shape, and the filling material is made to be small inside and large outside to prevent invagination under external force.

填补天柱位置圆孔时，适当将孔洞和填补材料打磨成不规则喇叭口形状，防止多年以后出现转动脱落。

When filling the round hole at the sky pillar, properly polish the hole and filling material into an irregular bell mouth shape to prevent rotation and falling off after years of use.

填补以后的效果。

This is the effect after filling.

古琴形制在古代通常有严格的规范，但因历史上的人口流动、信息传播等受到当时科学技术发展的制约等原因，斫琴技艺仅流传于少数宫廷和宗派斫琴师之间，所以民间常出现不符合规制的“野斫”古琴作品，并流传至今。

这张琴底板上五个大小不一的开孔，属于不合规制的臆造，对琴的音色有较大的不良影响，为了修好以后有良好的音色，修复中必须进行填补。填补材料的配备，跟缺损修复法一样，寻找相同或者相近的木材，用纸和铅笔画出孔洞大小和形状的纸样，根据纸样裁好填补材料，填补材料打磨成稍微内小外大形状，填补的时候用铲刀和锉将孔洞口也打磨成喇叭状。这样处理，一是好施工，二是利于粘接，三是可以防止以后修补材料向内塌陷。

填补正圆形孔洞时，不但要防止内陷处理，还需要将圆形处理成非正圆形，防止多年以后出现转动、脱落。孔洞和填补材料要做好方向标记，以免因方向错误而出现大小不合或打磨过度的情况。

The shape of Guqin was subject to strict norms in feudal society. Yet due to restrictions on human flow and information dissemination caused by backward technological development in history, the skill of Guqin making was only possessed by a few royal and sectarian Guqin masters, hence the emergence of "folk-made Guqin" works that did not conform to the regulations ,which have been handed down to this day.

The five holes of different sizes on the bottom plate of this qin are unregulated fabrications, having great adverse effects on its timbre. To restore the timbre, these holes must be filled in the process of repairing. The provision of filling materials is the same as the process of defect repair: looking for the same or similar wood; drawing the pattern of the holes on the paper, displaying the size and shape of holes; cutting the filling material according to the pattern, polishing the filling material to make it to be small inside and large outside; polishing the holes into bell mouth shape with a shovel knife and file during filling. This method can be easily operated, make bonding easier, and prevent the repair material from collapsing inward in the future.

When filling a circular hole, polish it into a non-right circle shape to prevent it from sinking inside and rotating and falling off after years of use. The direction of holes and filling materials should be marked to avoid size inconsistency or excessive polishing due to the wrong direction.

第五章　配件修备法

Chapter V
Guqin Accessories
Repair Method

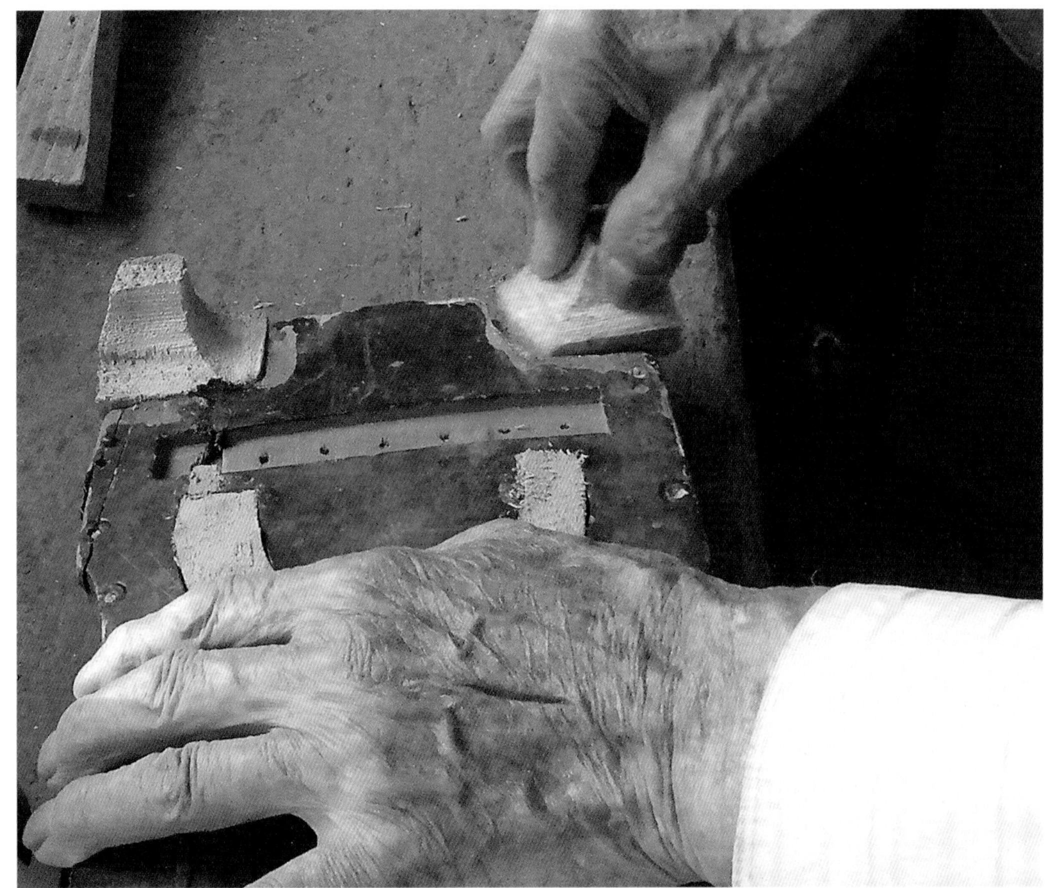

根据所修缮古琴的形制，参照其历史原貌或同时代相近风格的古琴，制作相匹配的配件。

Make accessories for the Guqin to be repaired according to its shape and original appearance or qins with similar styles of the same generation.

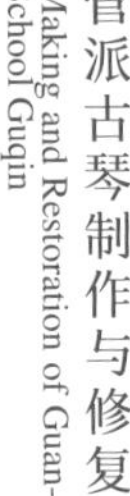

安装配件时除了剔除琴体腐朽木质，尽量不要改变琴体原貌，多打磨配件，让配件适应琴体。

When installing accessories, remove the decadent wood from the qin body, be careful not to damage the original qin body, polish the accessories more to make them fit the qin body.

古代老琴配件缺失是古琴修复中最常见的问题，如果单侧缺失，修复就比较简单，参照另一侧形状制作复原即可。如果双侧均缺失，则需要寻找曾经的照片、文字记录、当事人口述等资料，按照历史信息配备制作。如果无资料信息，则根据老琴制作时代的审美标准，参照同时代相同形制的配件制作方式，设计与之相匹配的配件。

这张琴除岳山以外，其余配件均已缺失，但从岳山可以看出配件所用的材质。由于此琴属于历史“野斫”，故以参考同时代相同形制古琴配件为主，琴体留下的配件痕迹为辅的方式进行修复。在配件装配中，除了剔除腐朽木质，尽量不要伤及琴体原貌，多打磨配件，让配件适应琴体。

The most commonly seen problems in Guqin restoration is the lack of accessories. If one side is missing, the restoration is relatively simple--just make that of the other side according to the shape of this side. If both sides are missing, it is necessary to look for historical information such as photos, written records, oral statements of related master makers. If no information is available, design the accessories based on the aesthetic standards of the time when the qin was made and referring to the making methods of accessories of qin of the same shape.

Except the bridge, all the accessories of this qin are missing, but we can tell the materials used from the bridge. Since this qin belongs to the historical "folk-made Guqin", it is repaired by referring mainly to the accessories of the qin of the same shape and the same generation, and the traces of accessories left at the qin body. When installing accessories, remove the decadent wood, be careful not to damage the original qin body, polish the accessories more to make them fit the qin body.

第六章 漆胎修复法（上）

Chapter VI Lacquer Cement Repair Method (I)

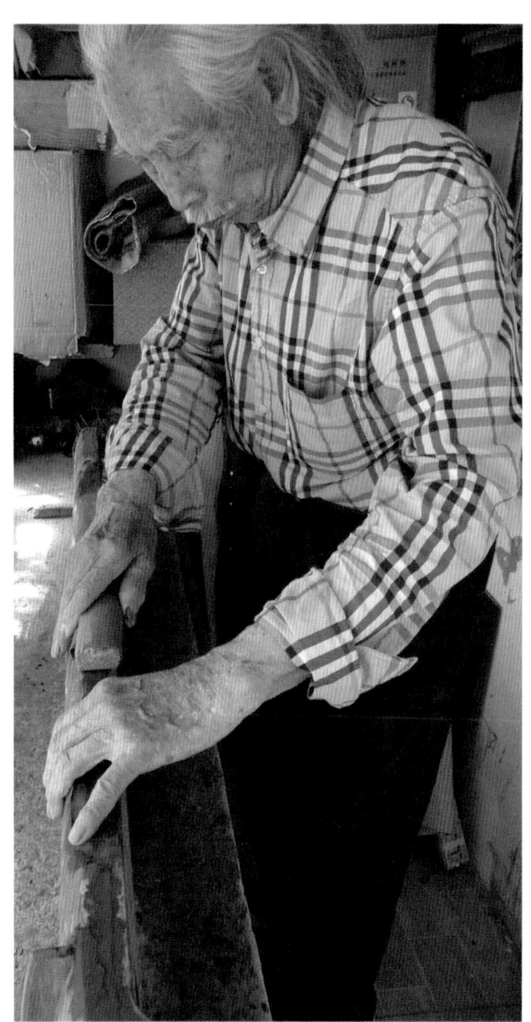

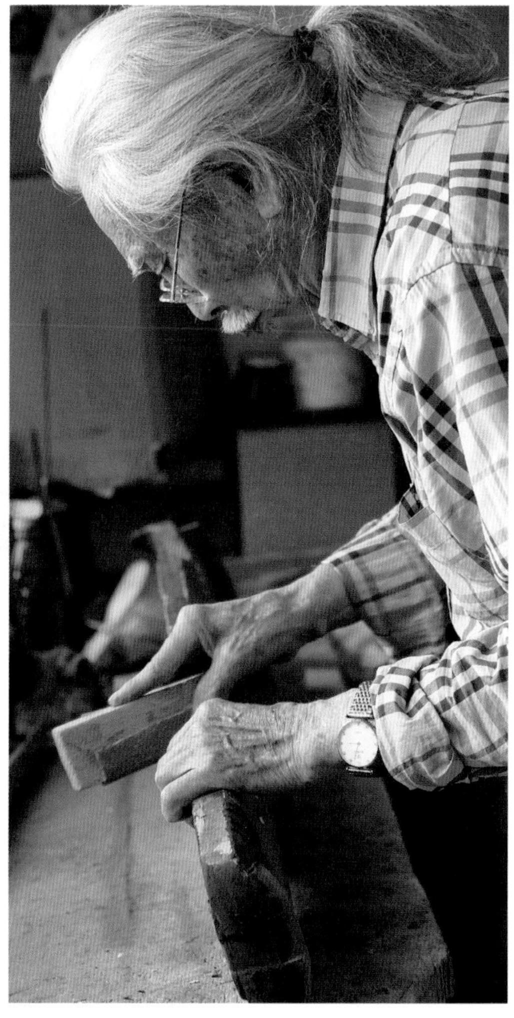

将琴体腐朽、凹陷、开裂等修复的部位进行精细打磨，注意保护琴体原有的老漆面，不可磨损或划伤，以便完工以后未修缮部位保持原貌。

Finely polish the decayed, sunken, cracked and other parts to be repaired of the qin body, while protecting the original paint surface of the qin body from abrasion or scratch, so as to keep its original appearance after reparation.

在打磨中发现琴体近段时间出现了严重弓背样变形，这与此琴无灰胎，长期不弹奏，木质部分腐朽、过度干燥，修补灰胎以后放入湿度较高的阴房有直接关系，必须进行矫形，修复方案也需重新调整。

During polishing, it was found that the qin body has serious bow-back deformation in recent days, which was directly related to the lack of lacquer cement, long-term idle after use, partly decayed wood and overly dry environment, and being put into a shady room with high humidity after repairing its lacquer cement. It is therefore must be corrected and the repair scheme needs to be readjusted.

完成打磨后，对琴体进行矫形。

Correct the qin body after finishing polishing.

一张老琴的漆面断纹是一种年代的记忆，是时间在琴面刻画下的年轮，它是除了腹款和琴底篆刻文字之外，能够推断老琴斫制年代的直接依据。修复一张能够再次演奏的老琴，在确保实用性以外，首要的就是保护好原有的漆面和断纹。

老琴常见小问题就是断纹翘起和脱落，弦路位置的翘起会导致弹奏中出现煞音，只需将翘起部分轻轻打磨平即可。如果翘起严重，甚至出现脱落，使用生漆调配适量酒精后用注射器注入，注入时一定要把握好漆量，过多会溢出，过少则又达不到黏合效果，注入后压紧等待干燥再打磨，切不可过度打磨或再施以灰胎、面漆等。弦路以外轻微翘起无须修复，如有严重翘起或出现脱落与将要脱落的情况，则使用前边方法注入生漆黏合，干燥以后无须打磨。

此琴无灰胎，仅薄层漆面。虽为清代所斫，但因过薄漆面无法形成断纹，琴额、面、尾漆层损伤严重，导致部分木质腐朽。根据具体情况对症处理以后，进行了灰胎填平与修补。原计划保留原有的未损坏漆面，使修复以后外观充分呈现原来的历史信息。但在灰胎干燥后的打磨中，发现琴体呈严重弓背样变形，这与此琴无灰胎，长期不弹奏，木质部分腐朽、过度干燥，修补灰胎以后放入湿度较高的阴房有直接关系，必须进行矫形，修复方案也由之前的保留外观大部分历史信息变更为矫形以后重新制作灰胎。

The broken patterns on the surface of an aged qin carry a memory of several ages and are the annual rings depicted by time on the qin surface. It is an element that best tells the age of an aged qin, following the stamps on the qin belly and the engraved characters on the bottom of the qin. To repair an old qin that can be played again, besides ensuring its practicality, the first thing is to protect the original paint surface and broken patterns.

Warping and falling off of broken patterns are commonly seen in aged qins. If a bad sound is produced by the warping of the string path, polish the warped part slightly is enough. If the part is cocked seriously or even falls off, use raw lacquer to mix a proper amount of alcohol and inject the mixture into the hollow part with a syringe. During injection, the amount of lacquer must be well controlled. Too much lacquer may lead to overflow, while too little lacquer may fail to bond the cocked part. After injection, press it tightly and wait for drying before polishing. Do not over-polish or apply lacquer cement, topcoat, etc. Slight warping outside the string path does not need to be repaired. If there is serious warping or falling off or about to fall off, inject raw lacquer with the previous method for bonding, and no polishing is needed after drying.

This qin has no lacquer cement but only a thin layer of paint. Though made in the Qing Dynasty, this qin's paint surface was too thin to form broken patterns, and the paint layers on the forehead, surface and tail of the qin are seriously damaged, resulting in part wood decay. After symptomatic treatment according to the specific situation, fill in and repair the lacquer cement. The original undamaged paint surface was planned to be retained so that the appearance fully presents the original historical information after restoration. However, during polishing after the lacquer cement is dried, it is found that the qin body is seriously deformed like a bow back, which was directly caused by the lack of lacquer cement, long-term idle after use, partly decayed wood and overly dry environment, and being put into a shady room with high humidity after repairing its lacquer cement. It is therefore must be corrected and the repair scheme needs to be readjusted from retaining most of the historical information of the previous appearance to re-making the lacquer cement after shape righting.

第七章　琴体矫形法

Chapter VII Qin Body Shape Righting Method

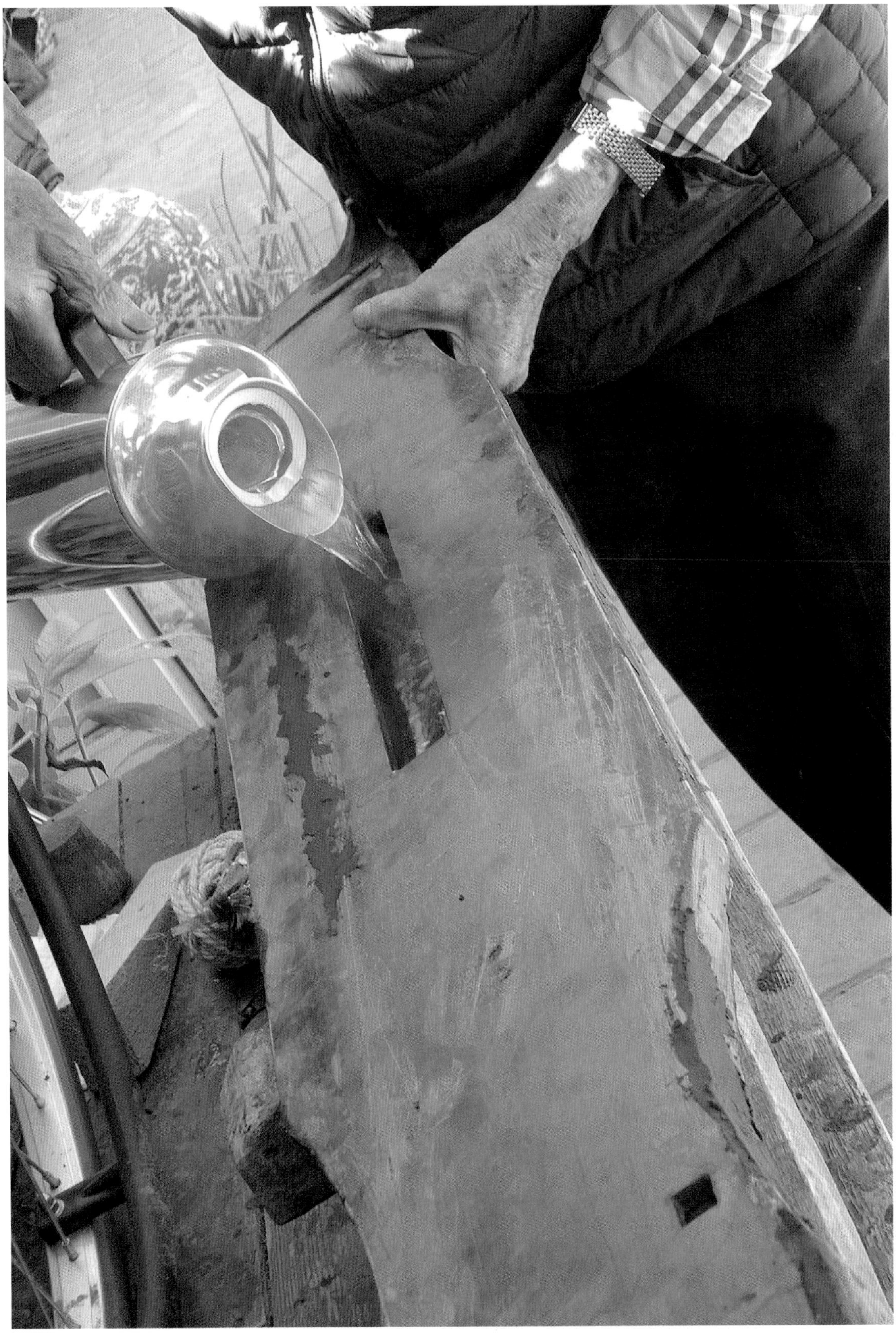

给琴腹内注入开水，摇晃均匀，停留3至5分钟后倒出。

Inject boiling water into the abdomen of the qin, shake evenly, let it stay for 3-5 minutes and pour it out.

将弧形枕垫于琴面变形处的最高点，用绳索将琴头部位固定在矫形板（顺直的厚木板）上。

Place the arc pillow at the highest point of the deformation of the qin surface, and fix the head of the qin on the orthopedic plate (straight plank) with ropes.

琴尾和矫形板用木工夹固定，适当拧紧，但此时不宜过紧，以免出现断裂。

Fix the qin tail and the orthopedic plate with woodworking clips and tighten them properly, otherwise, they may fracture.

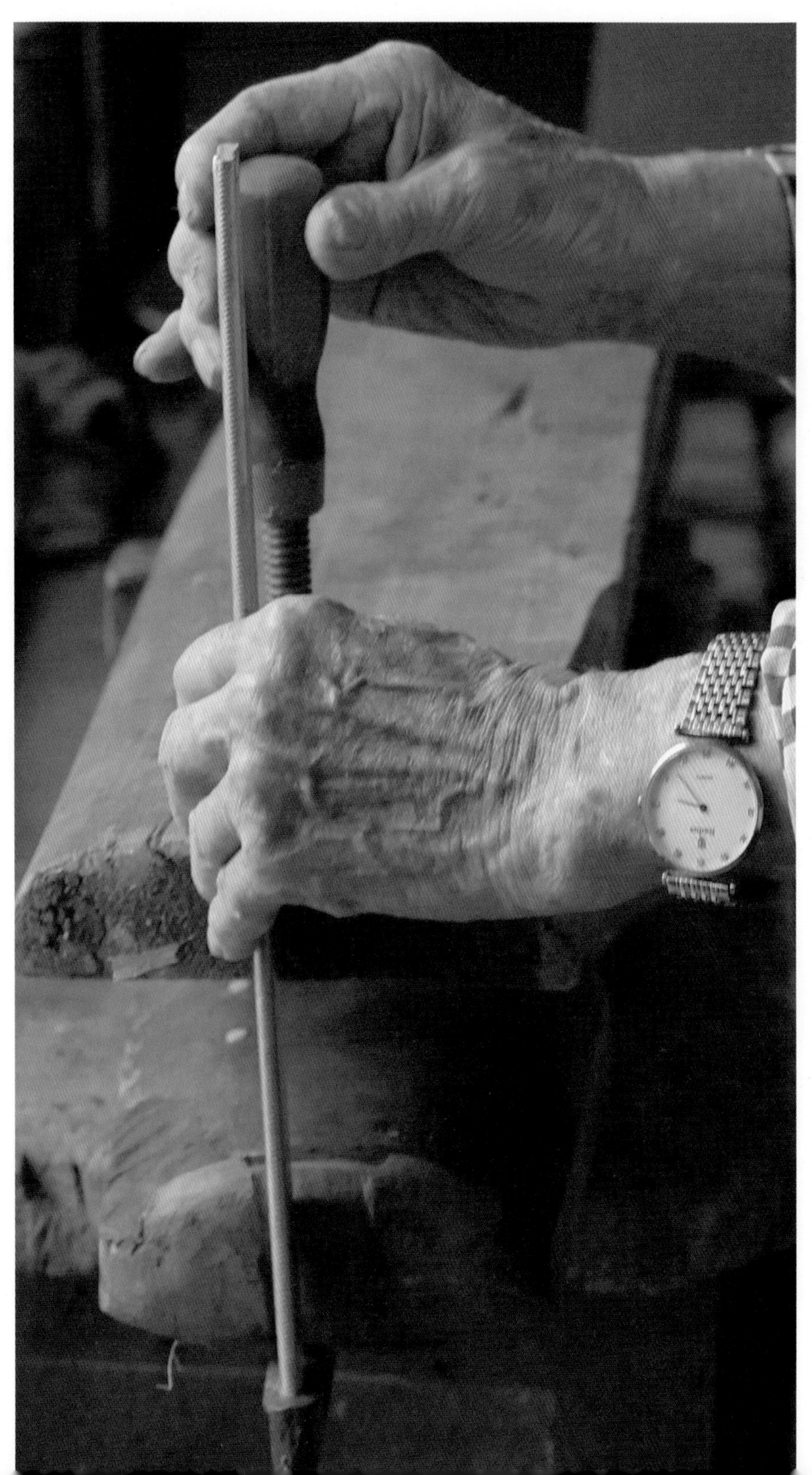

再次向琴腹内注入开水，注入量要比第一次有所增加，达到琴腹容量的 30% 至 50% 为宜。

Inject boiling water into the abdomen again, with increased amount, reaching 30%-50% of the abdomen capacity.

摇晃均匀，逐步而缓慢地一点一点拧紧木工夹。

Shake evenly, gradually and slowly tighten the woodworking clip bit by bit.

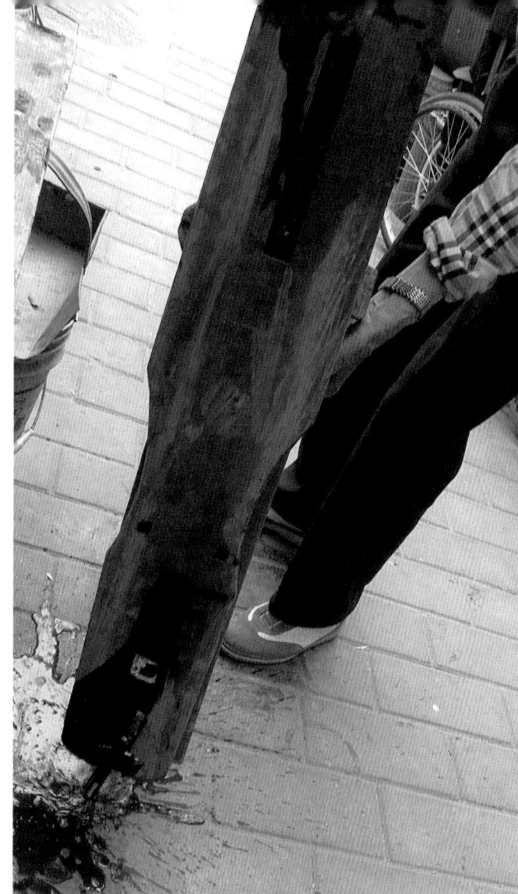

开水在琴腹内停留 3 至 5 分钟后倒出。

Let the boiling water remain in the abdomen of the qin for 3-5 minutes and then pour it out.

用眼睛观测是否矫形到位。

Observe with eyes whether the shape is rightly corrected.

用靠尺测量琴底是否平直，矫形平直到位以后，再适当拧紧木工夹，使其出现“矫必过”的情况，这样松开以后才能矫形到位。

Use a guiding rule to measure whether the bottom of the qin is straight or not. When it is straight, tighten the woodworking clips properly for excessive correction, so that when the clips are removed, the plate can remain straight.

再次用眼睛观测，“矫必过”程度不宜太浅，也不宜过深。

Observe with eyes whether the "correction" is excessively done, to a proper degree.

老琴变形也是常见问题，特别严重的还会出现扭曲，可使用捆绑法、火烤法、开水滚烫法等对症矫形。变形严重的必须分步骤多遍矫形，切不可急于求成，造成人为的第二次损坏。

琴友在家使用的琴也较常出现弓背或塌腰情况，导致打板和抗指，一般在家可使用重压矫形法进行缓慢矫形处理，出现弓背的在琴面变形高点放置沙袋或者其他重物，塌腰则将琴底朝上放置重物，经过一段时间的加压，观察是否恢复到位。未恢复到位的继续加压，直到恢复到位为止。

此琴历史相对久远，缺乏有效的漆胎保护，木质腐朽、过度干燥，长期未使用，矫形中极易出现破损、断裂等意外情况，工作中必须小心谨慎，提前制定万无一失的矫形方案。

本次方案最终确定使用开水滚烫法与捆绑法相结合的方式进行，因为开水注入琴内，可以有效缓解琴体木材因为腐朽干燥而出现的脆性，防止在捆绑复位的时候出现断裂。复位时使用“F”形木工夹，缓慢地一点一点复位，仔细观察琴体是否有出现破损、断裂的先兆，如果发现有破损、断裂的危险，及时停止并松开。矫形达到要求以后再适当“过”一点，预留松开以后的回弹空间，一般在阴凉干燥处放置 3 至 5 天即可。

开水注入琴内不宜停留时间太长，一般 3 至 5 分钟为宜，如果注入一次达不到效果可以倒出，再次注入滚烫开水，因为矫形中使用的主要还是开水的热量，而非浸泡作用。

Deformation is also a common problem for aged qins, and distortions may appear on ones with especially serious problems. Such qins can be corrected by binding method, fire baking method, boiling water boiling method, etc. Serious deformation must be corrected step by step and for multiple times, and rush for success may cause man-made secondary damages.

Deformations such as bow back or collapse waist are often seen in qins played by amateurs at home, resulting in board hitting and finger resisting. The heavy pressure orthopedic method can be used for shape righting at home. For instance, place a sandbag or other heavy objects at the deformation part of the qin surface in case of bow back; and when the waist collapses, place the qin bottom upward and put some heavy objects on it. After a period of pressurization, observe whether it is restored to the right place. If not, continue to pressurize until it is restored to the right place.

Due to a relatively long history, lack of effective protection on the lacquer cement, and long-term idle, this qin's wood is excessively dry and decayed. Accidents such as breakage and fracture are easy to occur during shape righting. Work out a scheme in advance and be careful in the restoration process.

The scheme is determined as a combination of the boiling water method and binding method. Because the injection of boiling water into the qin can effectively relieve the brittleness of the wood of the qin body due to decay and dryness, and prevent fracture during binding and reset. When resetting, use the "F" shaped woodworking clips. Observe carefully whether there is any sign of damage or fracture on the qin body, if so, stop and loosen it in time. After the treatment meets the requirements, continue to apply pressure for excessive correction, thereby leaving room for the rebound after loosening. Then place it in a cool and dry place for 3-5 days.

Keep the boiling water in the qin for 3-8 minutes, but no any longer. If the effect cannot be achieved once, inject boiling water again, as it is the heat of boiling water that works in this process, not soaking.

第八章　漆胎修复法（下）

Chapter VIII Lacquer Cement Repair Method (II)

放弃以前“修旧如旧”的方案，调整为以演奏实用为主，补上一层灰胎可以起到保护琴体的作用，防止开裂变形。

Abandon the previous scheme of "repairing the old as before", and adjust to one that focuses on practical performance by adding a layer of lacquer cement to protect the qin body and prevent it from cracking and deformation.

在矫形中再次出现了面底板脱胶、开裂的情况，但是竹钉依然完好，所以只需要用灰胎填平即可。

During the shape righting process, degumming and cracking of the surface and bottom plates occur again, but the bamboo nails are intact, so we only need to fill and level it up with lacquer cement.

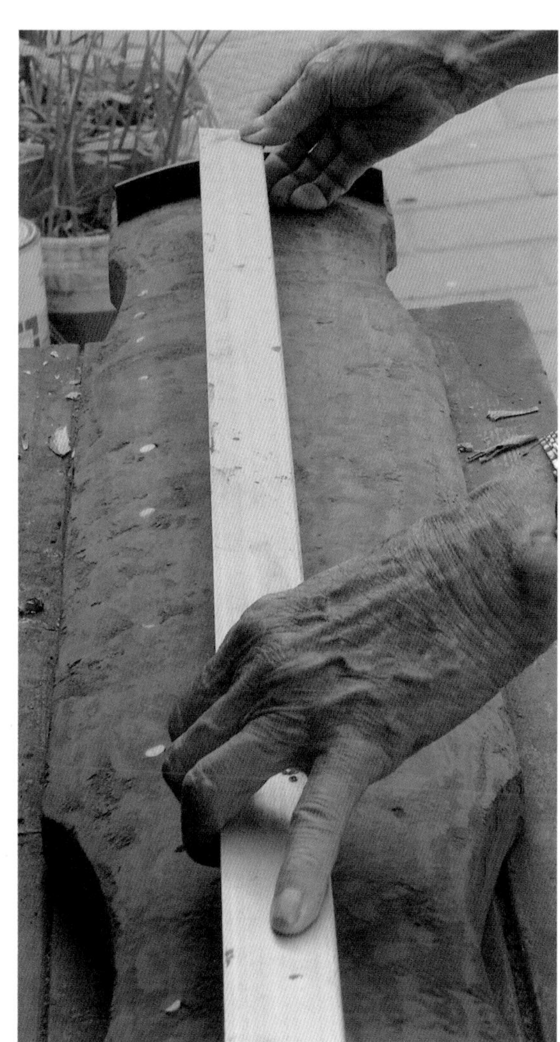

再次上灰胎、打磨、髹表漆的方法参照古琴制作技艺章节即可。

For second-time lacquer cement applying, polishing and surface painting, refer to the chapter of Guqin Making Skills.

以演奏实用为主的老琴修复，在兼顾音色的情况下，“修旧如旧”仍然为第一原则，万不得已不可破坏原有的历史信息。

此琴综合分析，考虑重新上灰胎、重新做面漆是因为这样既可以保护已经腐朽的琴体，防止它继续腐朽、变形、开裂甚至断裂，又可以让这张原本无灰胎的老琴音色更好。

灰胎、漆面如何制作，前边古琴制作章节已经详细介绍，不再赘述。做灰胎前必须对原有的漆面进行打磨、拉毛，这是为了让新的灰胎跟原有的老漆面紧密结合，如有必要也可以再刷一遍生漆，等待干燥以后再做灰胎。

For practical performance-oriented Guqin restoration, while ensuring good timber, the first principle of "repairing the old as before" still works, and destroying the original historical information should be regarded as a last resort.

Comprehensive analysis on this qin suggests the re-applying of the lacquer cement and topcoat, because this can not only protect the qin body from continuing to decay, deform, crack or even break, but also improve the timbre due to lack of the lacquer cement.

As we have introduced how to make lacquer cement and paint surface in detail in previous chapters on Guqin making, they will not be described in this chapter. Before making the lacquer cement, the original paint surface must be polished and napped to make the new lacquer cement closely fits with the original paint surface. If necessary, apply another layer of raw lacquer, and make the lacquer cement after it is dried.

第九章　修复成琴

Chapter IX Completion of Guqin Restoration

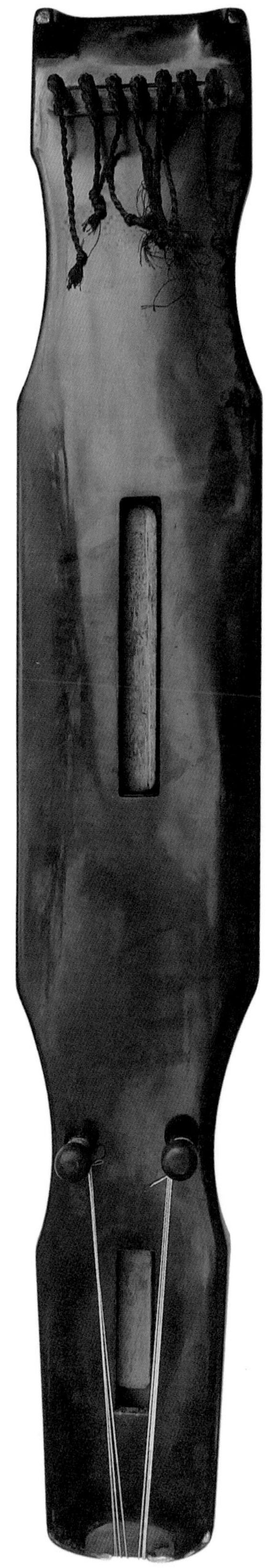

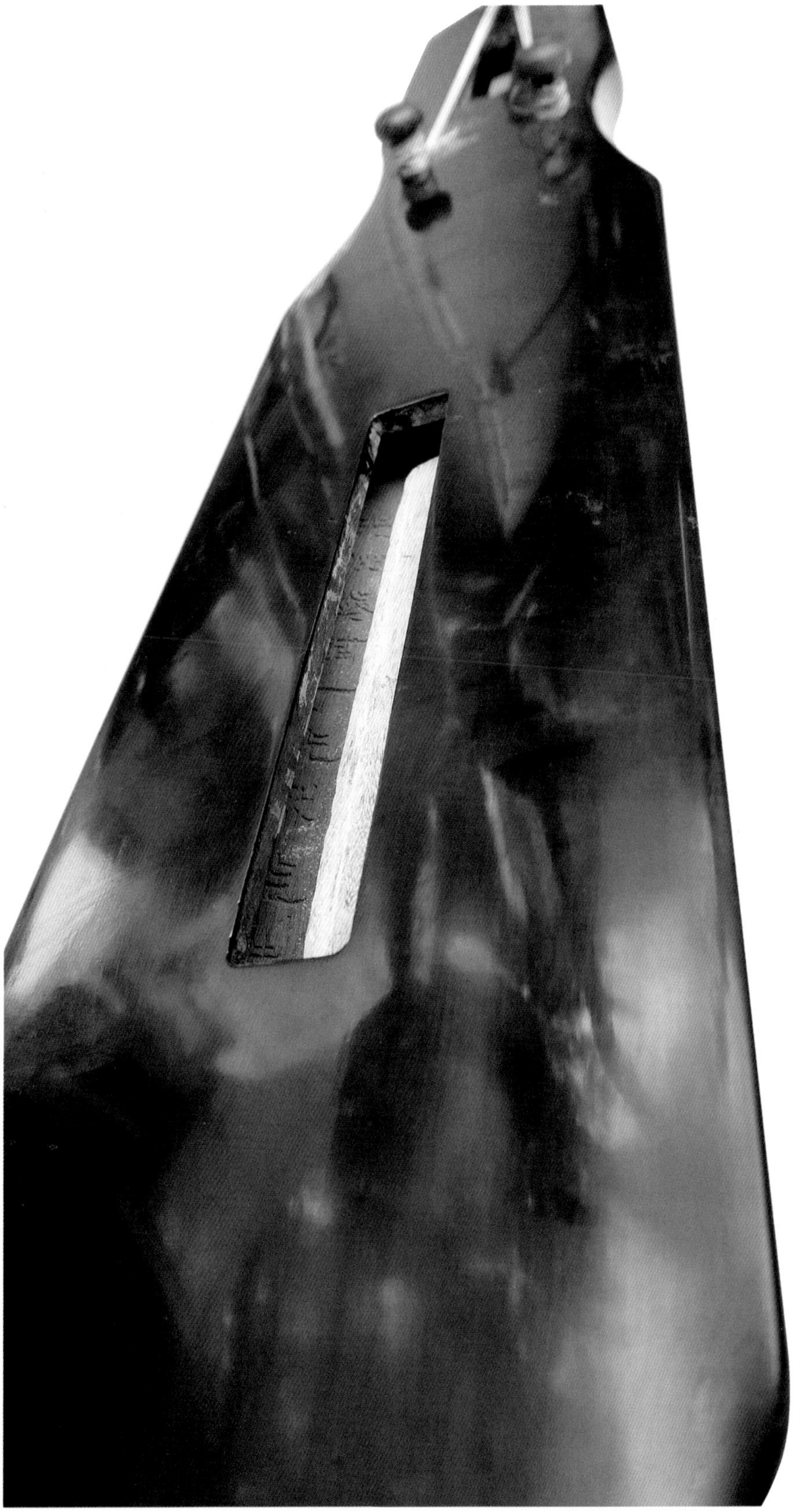

第十章 作品展示

Exhibition of Works

田双坤斫“鸣金玉震”琴 落霞式

通长124厘米，有效弦长112.4厘米，额宽19.5厘米，肩宽21厘米。

“鸣金玉震”琴，杉木面板，梓木底板，圆形龙池，方形凤沼，通体黑漆，乌木配件，琴额镶嵌长方形白玉，琴底圆形龙池下刻名“鸣金玉震”，下刻“田双琨之印”一方。

"Ming Jin Yu Zhen" Qin made by Tian Shuangkun, Sunset Style

Overall length 124 cm, effective chord length 112.4 cm, forehead width 19.5 cm, and shoulder width 21 cm.

The "Ming Jin Yu Zhen" Qin's surface panel is made of Chinese fur, and bottom plate catalpa. It has a round Longchi and square Fengzhao. The whole body is covered by black paint, supplemented with ebony accessories. A rectangular white jade is inlaid on the forehead, and the four Chinese characters "Ming Jin Yu Zhen" are engraved under the round Longchi at the bottom panel, together with "Tian Shuangkun's Seal".

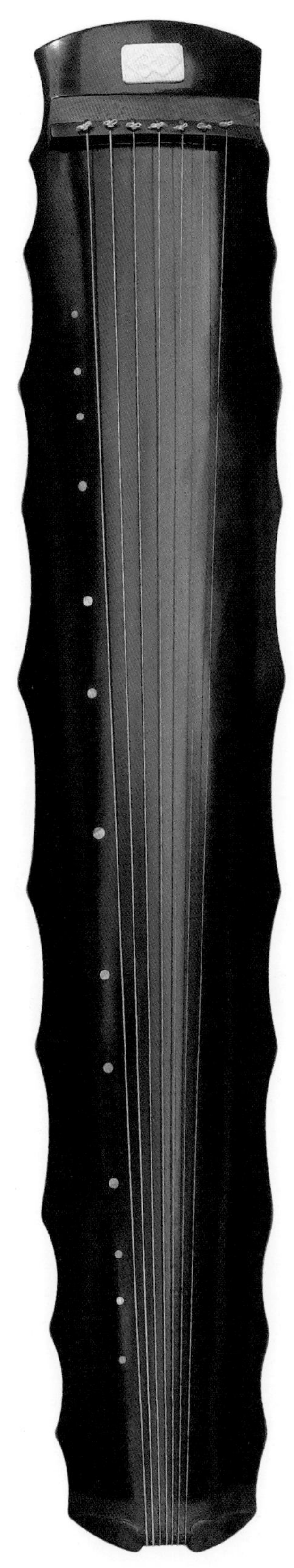

田双坤斫“明珠”琴 仲尼式

通长 124 厘米，有效弦长 113 厘米，额宽 19 厘米，肩宽 20 厘米。

“明珠”琴，杉木面板，梓木底板，面板栗壳色，底板黑漆，乌木配件，琴底刻名“明珠”，龙池两侧刻“春暖观鱼跃，秋高听鹿鸣”，龙池下刻“田双琨之印”“终极得道”两方大印。

"Pearl" Qin made by Tian Shuangkun, Zhong Ni Style

Overall length 124 cm, effective chord length 113 cm, forehead width 19.5 cm, and shoulder width 21 cm.

The "Pearl" Qin's surface panel is made of Chinese fur, and bottom plate catalpa, and the former displays a chestnut shell color, while the latter is covered by black paint. It is supplemented with ebony accessories. The bottom is engraved with its name "Pearl", and both sides of Longchi with the line "Watch the fish leaping in the warm spring, listen to the deer singing in the autumn". Under Longchi, there are two seals, one is of Mr. Tian Shuangkun and the other reads "Ultimate Tao".

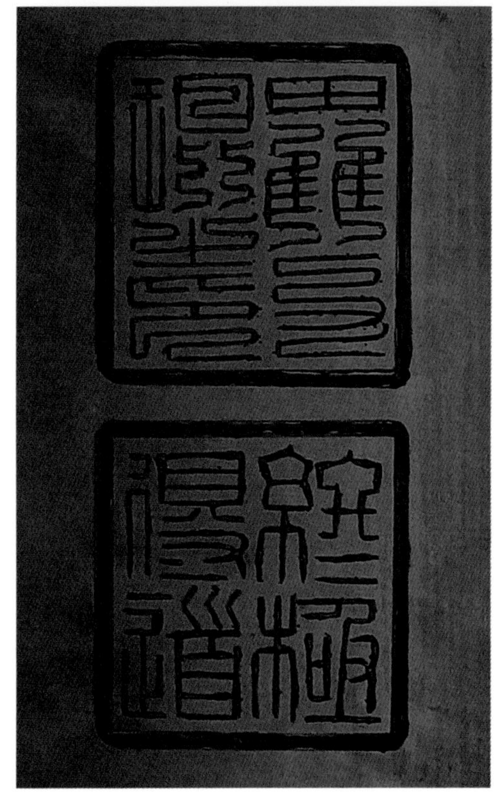

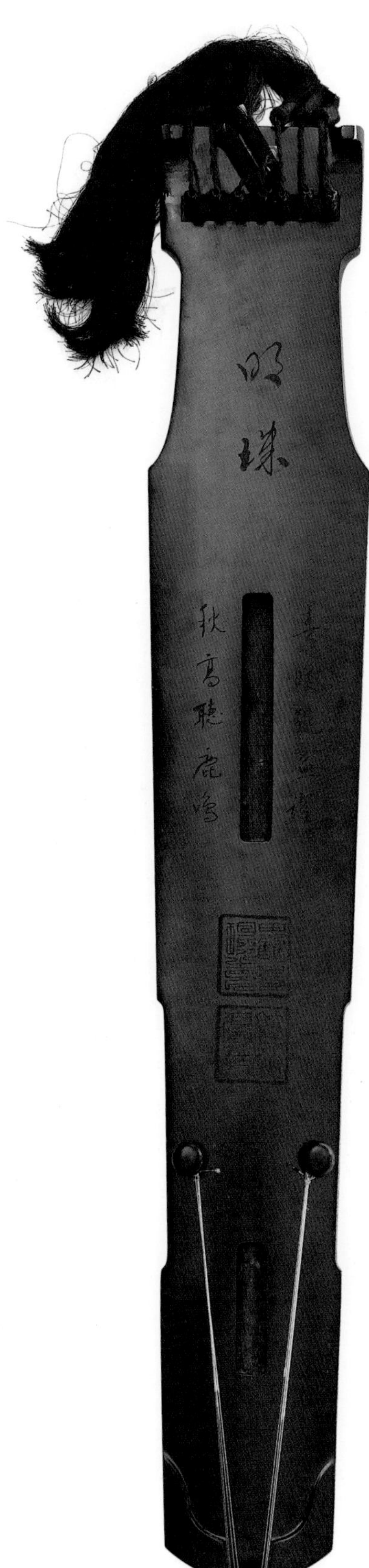
明珠
秋高聽鹿鳴

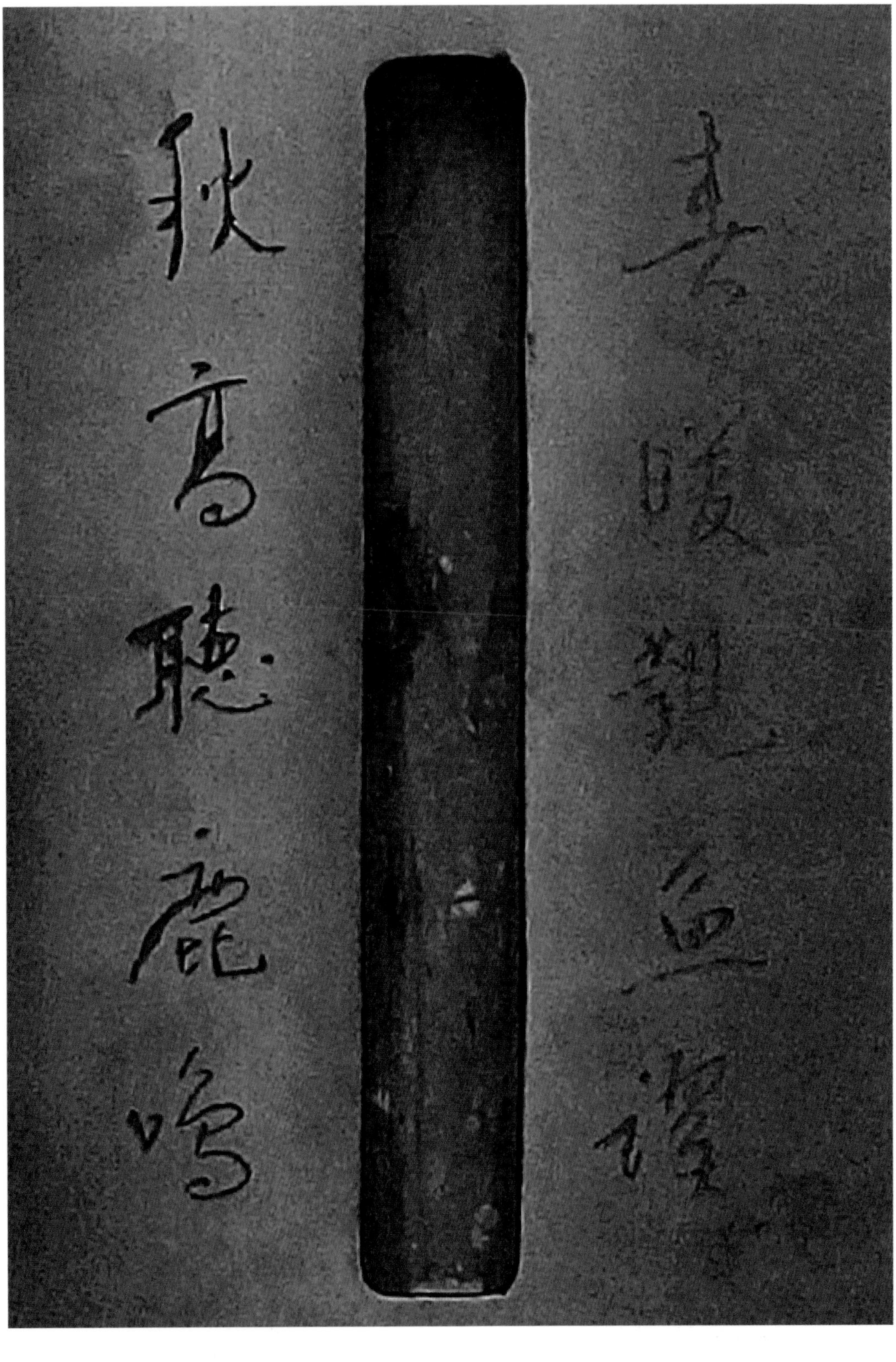

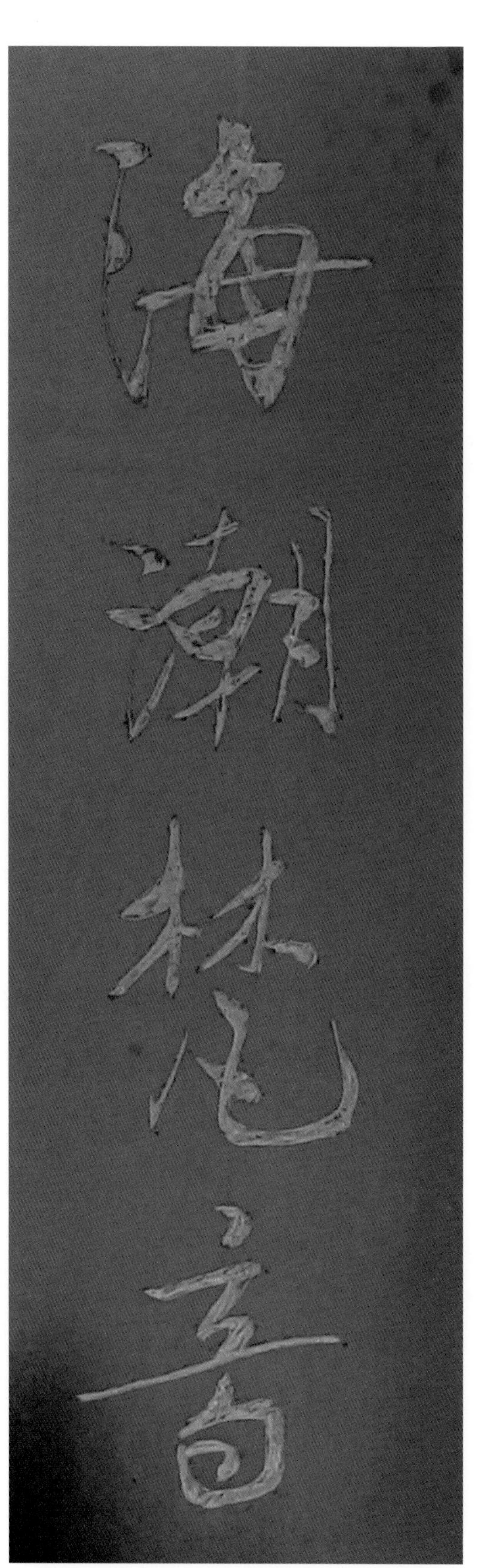

田双坤斫“海潮梵音”琴 混沌式

通长123厘米，有效弦长110厘米，额宽20厘米，肩宽21厘米。

“海潮梵音”琴，杉木面板，梓木底板，通体黑漆，乌木配件，琴底刻名“海潮梵音”，龙池下刻“田双琨之印”一方。

"Sea Tide & Sanskrit" Qin made by Tian Shuangkun, Taichi-chaos Style

Overall length 123 cm, effective chord length 110 cm, forehead width 20 cm, and shoulder width 21 cm.

The "Sea Tide & Sanskrit" Qin's surface panel is made of Chinese fur, and bottom plate catalpa. The whole body is covered by black paint, supplemented with ebony accessories. The qin name "Sea Tide & Sanskrit" is engraved on the bottom panel, while the "Seal of Tian Shuangkun" under the Longchi.

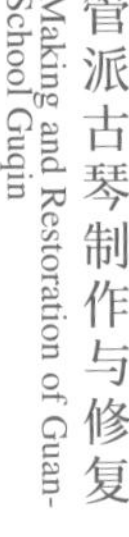

田双坤斫“悟”琴 连珠式

通长122厘米，有效弦长111厘米，额宽19厘米，肩宽20厘米。

“悟”琴，黄花梨素琴，琴底刻名“悟”，龙池下刻字为集板桥作品“难得糊涂”。

"Enlightenment" Qin made by Tian Shuangkun, Linked Bead Style

Overall length 122 cm, effective chord length 111 cm, forehead width 19 cm, and shoulder width 20 cm.

The "Enlightenment" Qin, a su qin made of fragrant rosewood, is engraved with the name "Enlightenment" at the bottom panel, and Zheng Banqiao's work "A Rare Confusion" below the Longchi.

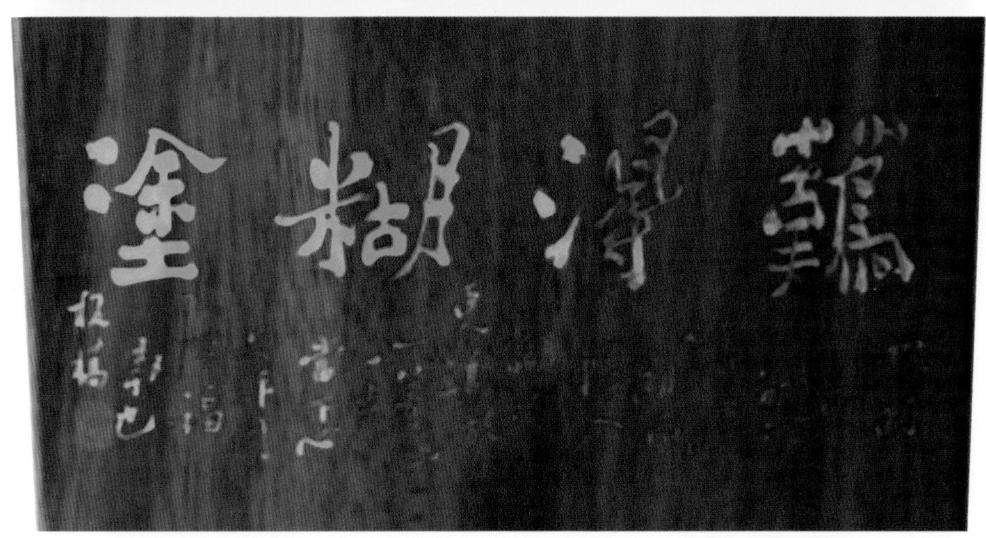

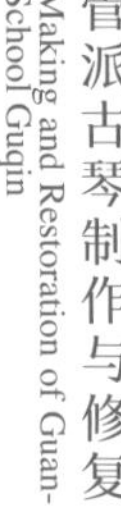

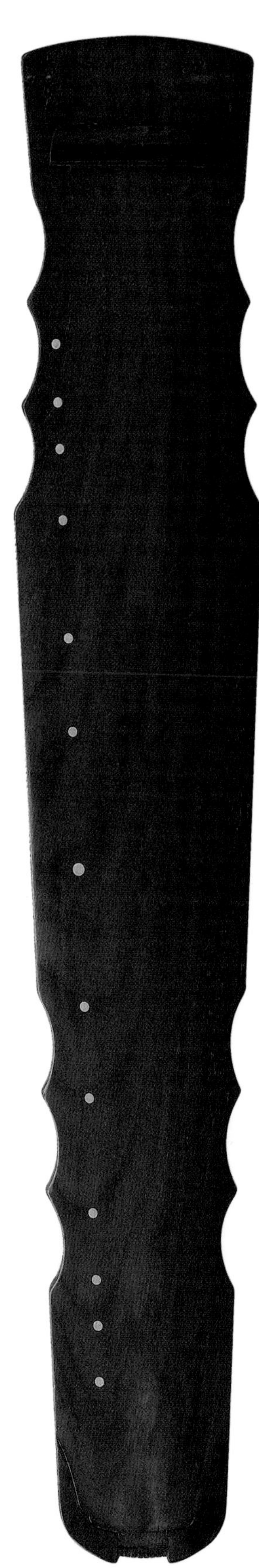

昭闻斫“幽玄”琴 神农式

通长121厘米，有效弦长115厘米，额宽20厘米，肩宽20厘米，尾宽14厘米，厚6.2厘米。

神农式“幽玄”琴，杉木百衲面板，楸木底板，通体黑漆，局部露鹿角霜灰胎，乌木配件，龙池、凤沼可见菱形百衲拼接。琴腹内墨书“乙未春昭闻斫，长安阿愚署”。琴底刻名“幽玄”二字由四川书法家洪厚甜先生题，龙池两侧由文学大家贾平凹先生书“得山水清气，极风云大观”，琴名下“昭闻制琴”和雁足间“古法斫琴”两方印章，由书法篆刻家韩春涛先生制。

"You Xuan" Qin made by ZhaoWen, Shen Nong Style.

Overall length 121 cm, effective chord length 115 cm, forehead width 20 cm, shoulder width 20 cm, tail width 14 cm, and thickness 6.2 cm.

The Shen Nong-style "You Xuan" Qin has a Chinese fur patchwork surface panel and catalpa wood bottom panel. The whole body is painted with black paint, and part of its cornu cervi degelatinatum lacquer cement is exposed. The accessories are made of ebony. The patchwork splicing can be seen in Longchi and Fengzhao parts. Inside the qin belly, it writes in ink "Made by ZhaoWen in the spring of 2015, subscribed by A Yu of Chang'an". The name "You Xuan" on the qin bottom was inscribed by Mr. Hong Houtian, a well-known calligrapher in Chengdu. The line "Drawing qi from the landscape to present magnificent views" by literary master Jia Pingwa is written on both sides of the Longchi. The two seals, "Qin Made by ZhaoWen", which is right below the qin name and "Guqin Making Technique" between YanZu were made by Mr. Han Chuntao, a calligrapher in Xi'an.

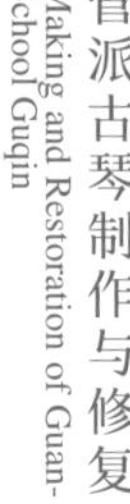

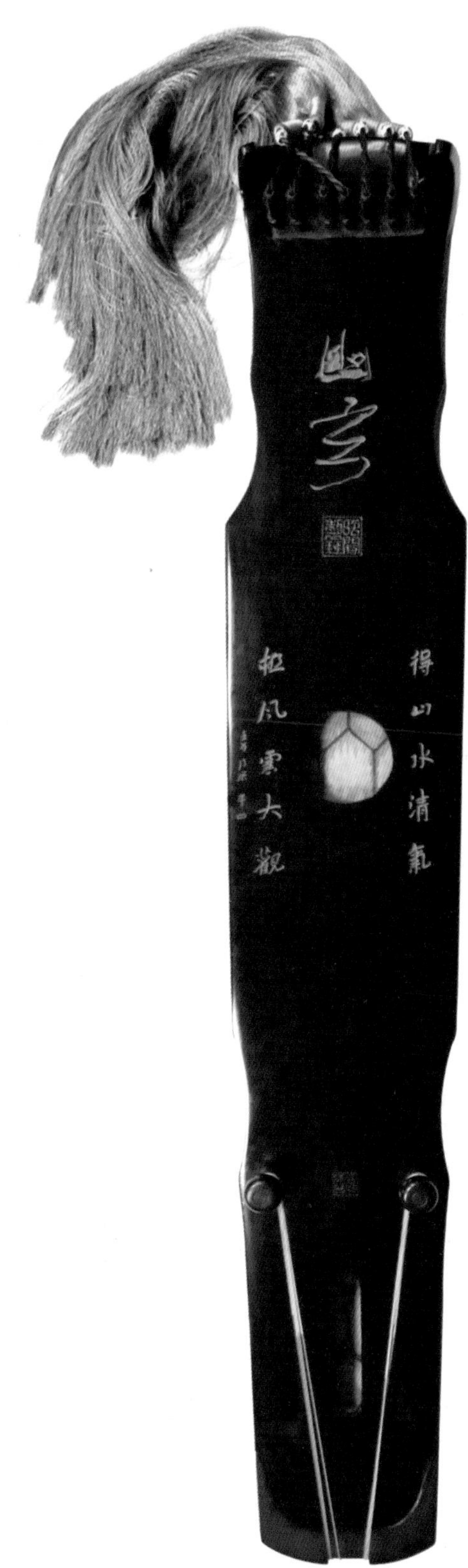
得山水清氣

得山水清氣
極風雲大觀

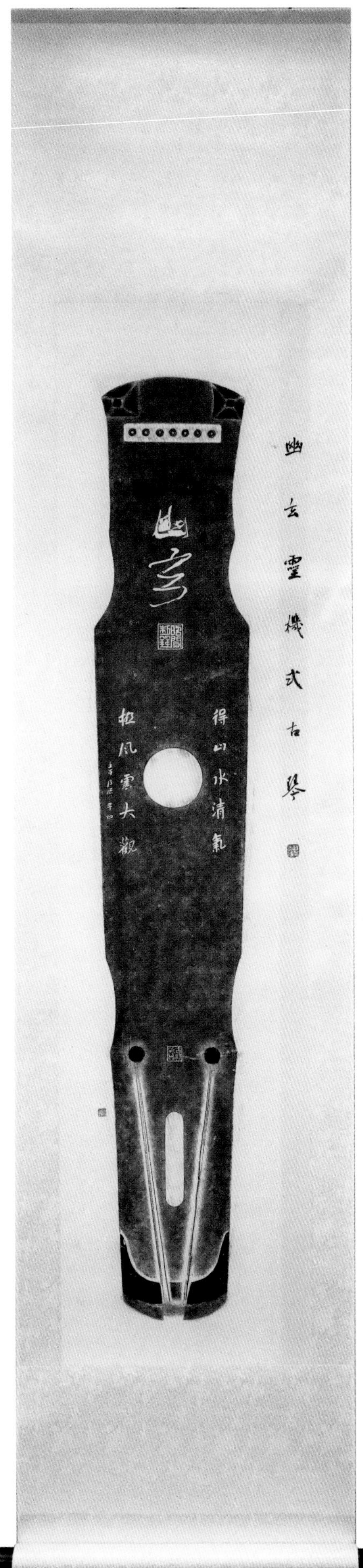
幽玄靈機式古琴
幽玄
得山水清氣

昭闻斫“镜花天”琴 洛象式

通长 120.5 厘米，有效弦长 105.5 厘米，额宽 21.5 厘米，肩宽 22.5 厘米，尾宽 15.5 厘米，厚 6.0 厘米。

洛象式“镜花天”琴，西安都城隍庙 660 年老松木房梁面板，香椿木底板，黑中飘金面漆，略透鹿角霜灰胎，乌木配件。琴腹内墨书“乙未秋昭闻斫于长安”，琴底刻大篆琴名“镜花天”。琴名下录杨宗稷《琴学丛书》语：“消忧之品，未有过于琴书者，琴以能怡情养性，延年益寿为功效，善弹琴者，借以炼性调气之用，非以悦他人之耳也，琴为有益身心性命之学道也，而非艺也。”落款“杨宗稷论琴，长安枫尚绿园主录”。落款小印两方，分别为“茹小石玺”“茹氏”。龙池下方有“河声岳色”长方形印章一枚，雁足之间绘金色云纹一处。

"Mirror, Flower and Sky" Qin made by ZhaoWen, Luoxiang Style

Overall length 120.5 cm, effective chord length 105.5 cm, forehead width 21.5 cm, shoulder width 22.5 cm, tail width 15.5 cm, and thickness 6.0 cm.

The Luoxiang-style "Mirror, Flower and Sky" Qin's surface panel is made of a 660-year-old pine beam from the Capital City God Temple, and its bottom panel is made of Toona sinensis wood. The finishing coat is mainly black, blended with gold. The cornu cervi degelatinatum lacquer cement is slightly exposed. The qin has ebony accessories. Inside the qin belly, it writes in ink "Made by ZhaoWen in the autumn of 2015 in Chang'an". The seal script name "Mirror, Flower and Sky" is engraved on the bottom. Below the name is a quote from the book Qinxue Congshu (Great Compendium of Qin Studies): "Guqin and books are the two best things to relieve melancholy. A qin can contribute to one's peace of mind and inner tranquility, and even prolong life. Those who are good at playing Guqin often rely on it to cultivate their mind, rather than pleasing others' ears. Guqin playing is a learning subject beneficial to people's physical and mental health, instead of being only an art form." The inscription "Yang Zongji talks about qin, recorded by lord of Chang'an Fengshang Green Garden" and two seals--"Ru Xiaoshi's Seal" and "Ru Shi" are below the quote. A rectangular seal reads "River Sound and Mountain Colors" is engraved below the Longchi, and a golden moire is painted between the YanZu.

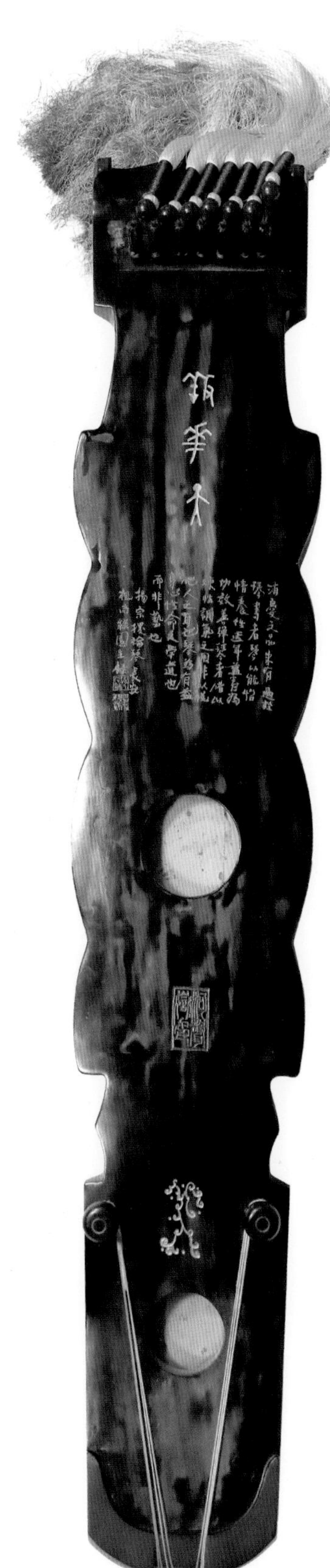

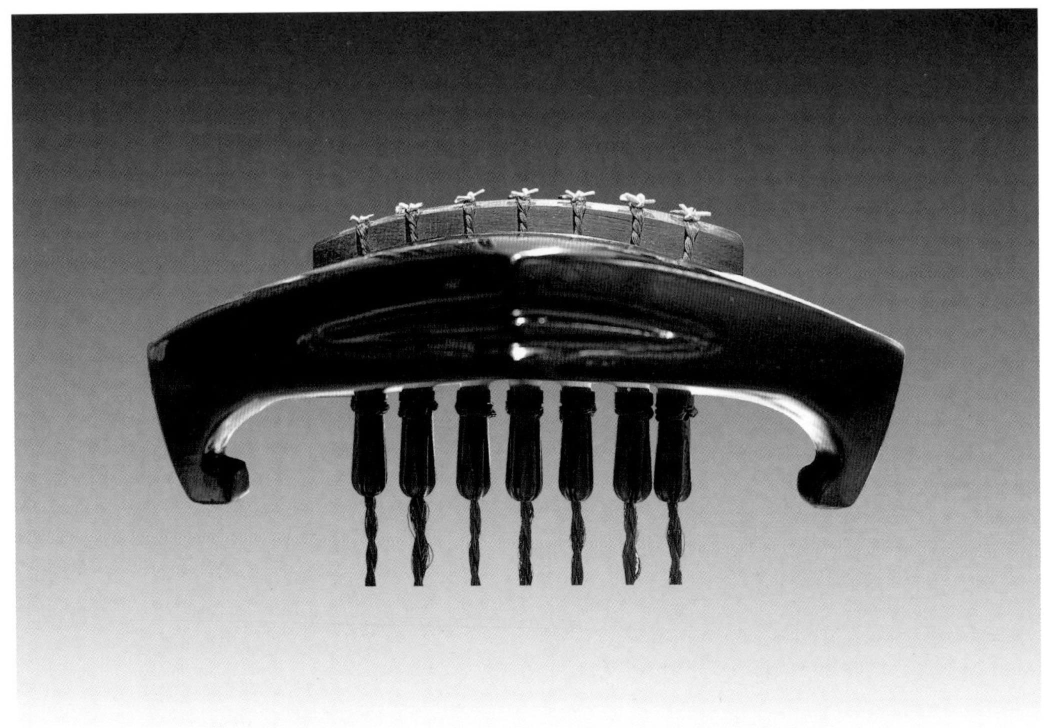

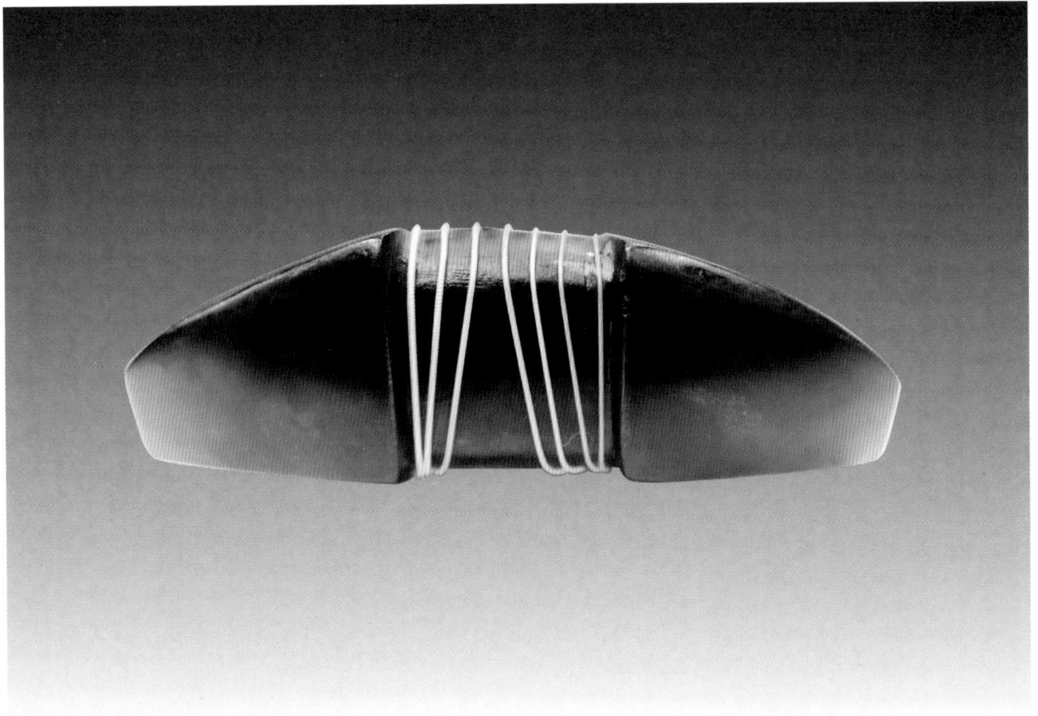

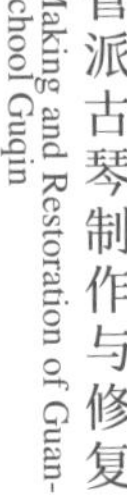

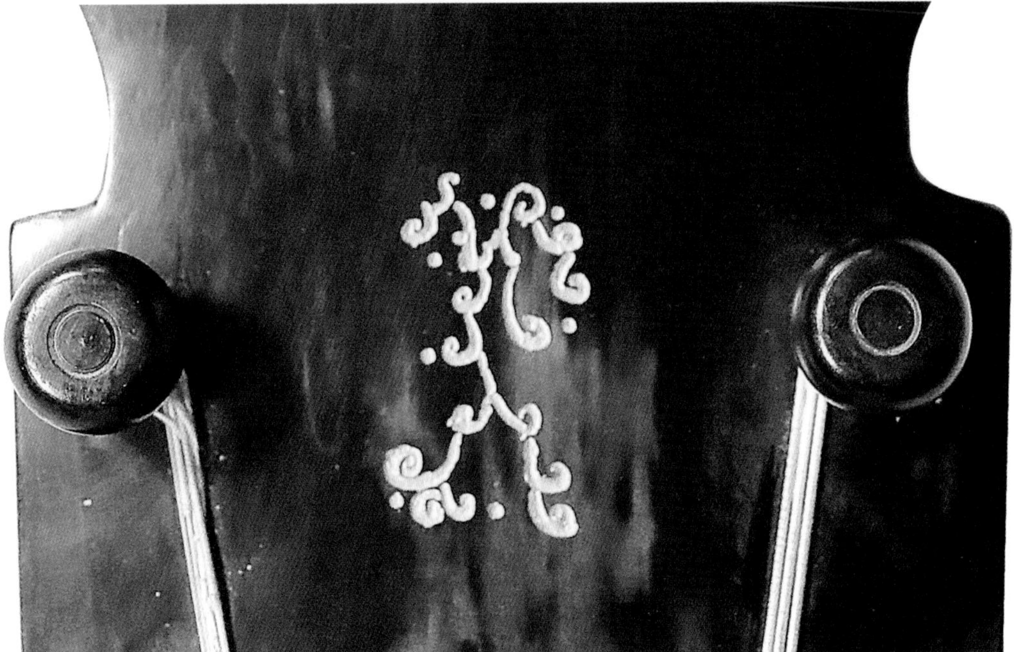

镜花天洛象式古琴

昭闻斫"燕子来"琴 神农式

通长121厘米，有效弦长108.8厘米，额宽20.5厘米，肩宽20.5厘米，尾宽13.8厘米，厚5.6厘米。

神农式"燕子来"百衲琴，西安都城隍庙660年老松木房梁百衲面板，香椿木底板，红黑相间面漆，略透鹿角霜灰胎，乌木配件，龙池、凤沼可见菱形百衲拼接。琴腹内墨书"乙未秋昭闻斫于长安，茗堂阿愚署"，琴底刻名"燕子来"，琴名下一方形"延年益寿"大印。龙池下录《琴学丛书》语："琴者纯粹为己，虽广坐操弄，亦必作独居，肆习时观，苟略有人之见存，则音随指变，不能成奏，欲求惬心，戛戛其难，盖未有气不可舒，而心能惬惬者。"落款"杨宗稷论琴语，三声堂主敬录"，落款小印"茹小石"。雁足间刻"自有乐地"方印一枚。

"Swallow Comes" Qin made by ZhaoWen, Shen Nong Style.

Overall length 121 cm, effective chord length 108.8 cm, forehead width 20.5 cm, shoulder width 20.5 cm, tail width 13.8 cm, and thickness 5.6 cm.

The Shen Nong-style "Swallow Comes" patchwork qin, whose surface panel is made of a 660-year-old pine beam from the Capital City God Temple in Xi'an and bottom panel of Toona sinensis wood, has red and black topcoat, with the cornu cervi degelatinatum lacquer cement slightly exposed. It has ebony accessories, and the patchwork splicing can be seen in Longchi and Fengzhao parts. Inside the qin belly, it writes in ink "Made by ZhaoWen in the autumn of 2015 in Chang'an, subscribed by A Yu of Mingtang". The bottom is inscribed with the name "Swallow Comes". Under it is a quote from the book Qinxue Congshu (Great Compendium of Qin Studies): "Guqin players play for themselves. Though sitting in a wide-open place, they should be deemed as staying alone. When a player practices, if someone stands nearby and offers some advice, he just does not know how to play, and the sound changes with his fingers. If he wants to be comfortable, his heart can be satisfied." The inscription "Yang Zongji talks about qin, recorded by Lord of Sansheng Tang", together with a small seal "Ru Xiaoshi" is under the lines. The YanZu part is engraved with a square seal of "Private Land of Pure Joy".

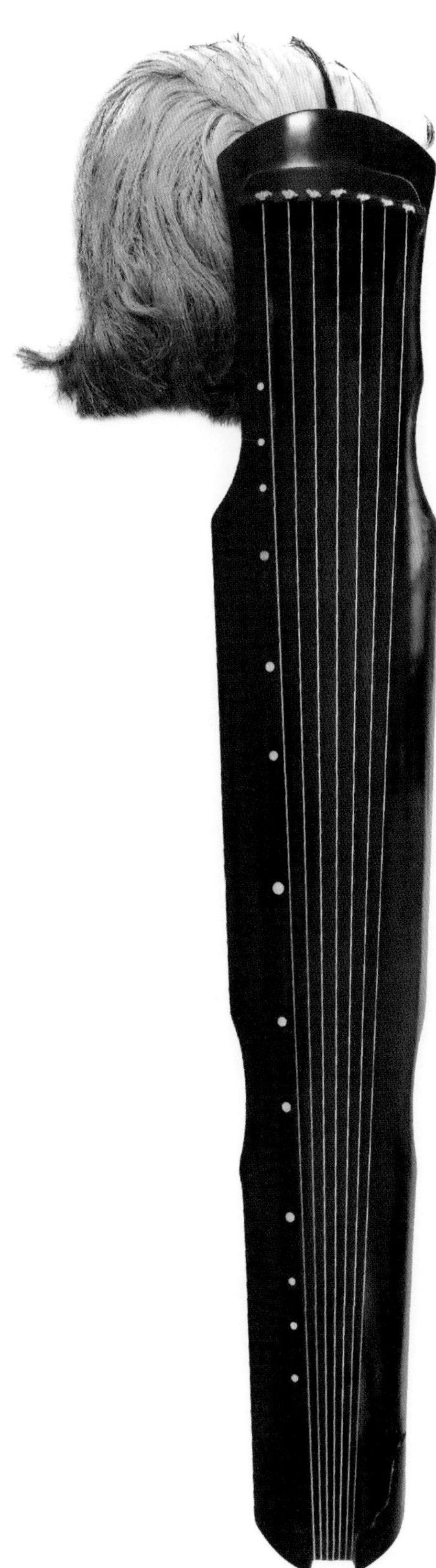

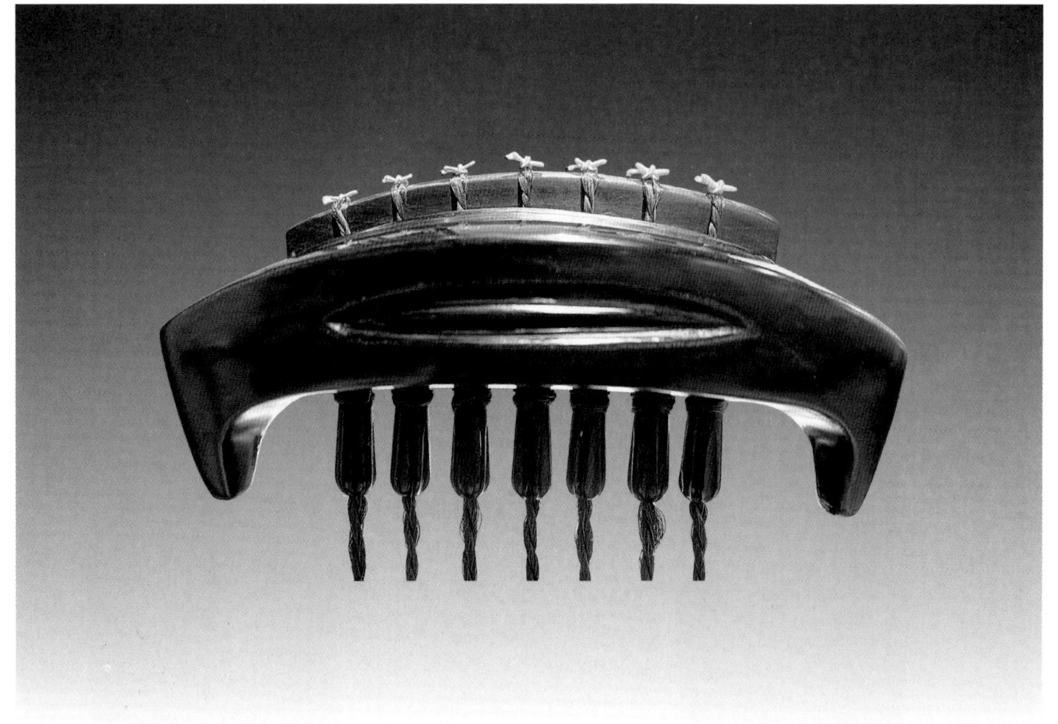

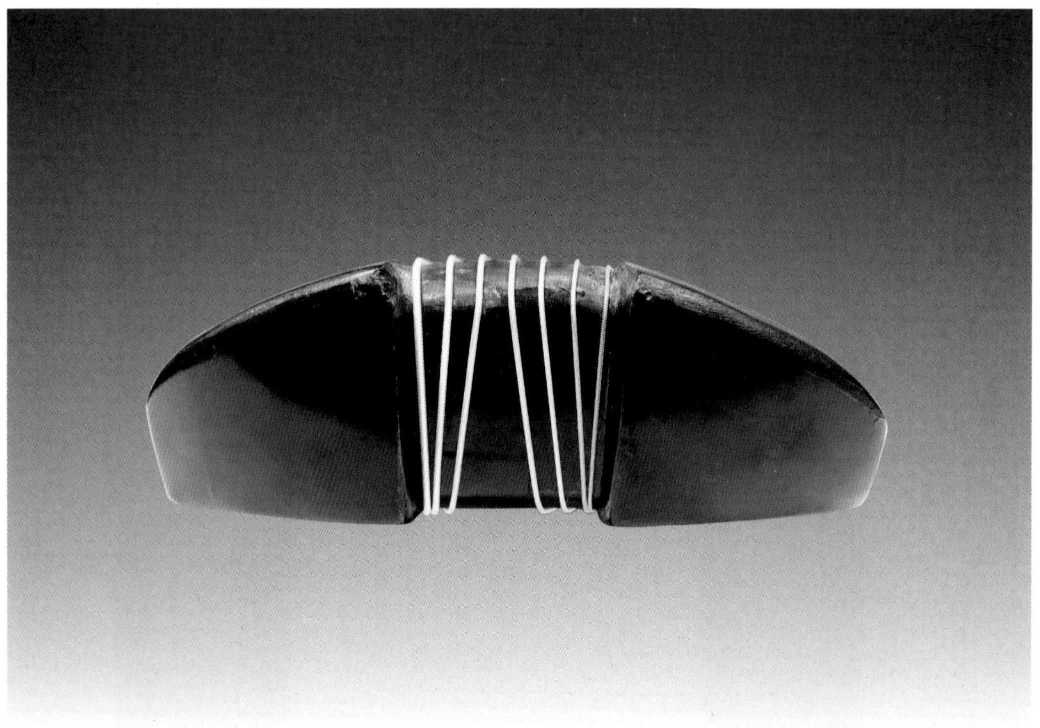

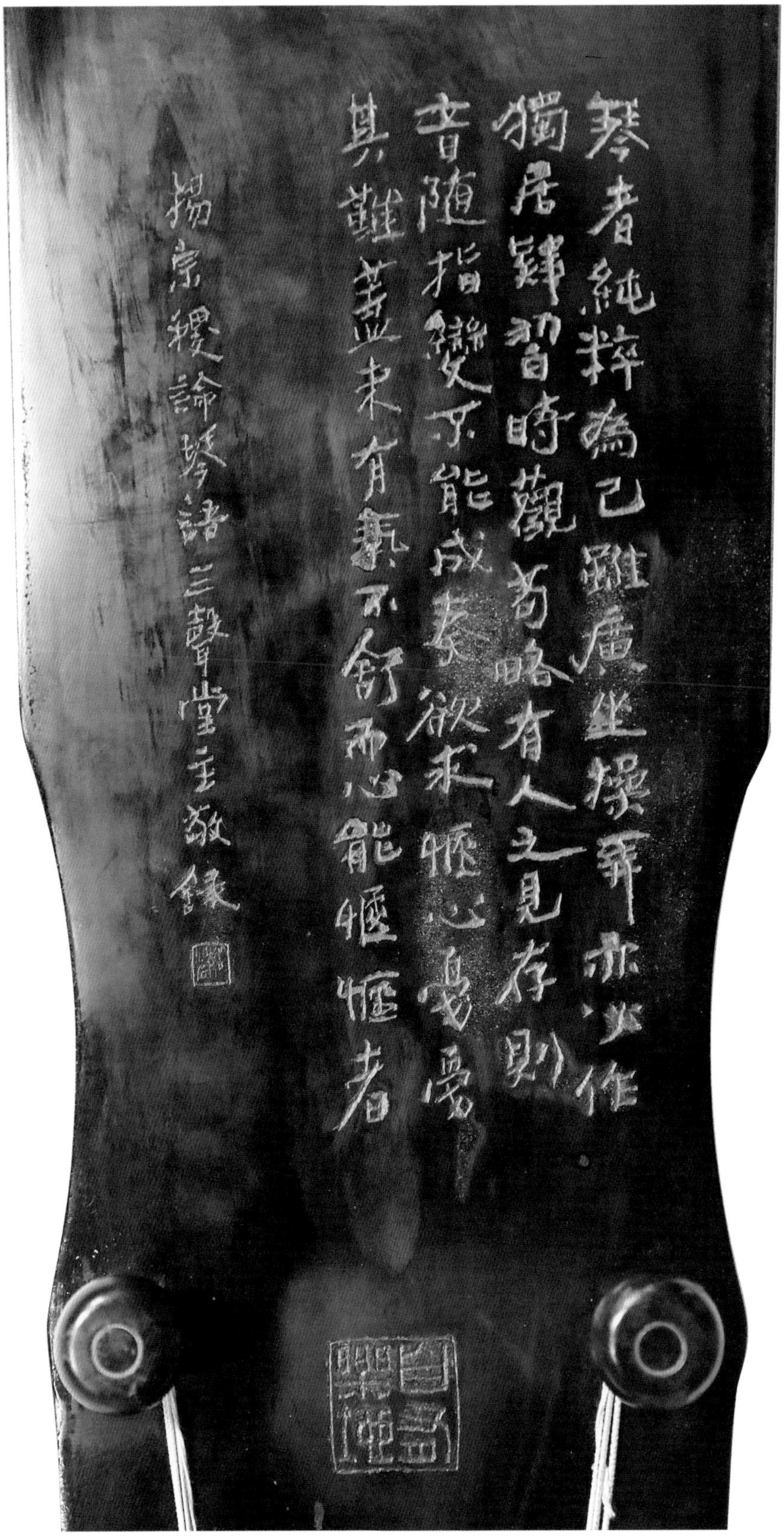

琴者純粹為己雖廣坐操弄亦必作
獨居肄習時觀苟略有人之見存則
音隨指變不能成奏欲求愜心愈憂
其難蓋未有氣不舒而心能愜愜者
楊宗稷論琴語 三聲堂主敬錄

燕子来 霊機式古琴

昭闻斫“寒潭月”琴 蕉叶式

通长124厘米，有效弦长111.6厘米，额宽18厘米，肩宽22厘米，尾宽15厘米，厚5.3厘米。

蕉叶式“寒潭月”琴，杉木面板，柏木底板，面漆红里透黑，局部可见鹿角霜灰胎，乌木配件。琴腹内墨书“乙未秋于广陵，昭闻手斫，茗堂阿愚署”，琴底刻名“寒潭月”三字由西安文博金石大家茹小石先生题，龙池两侧刻有集句“琴言清若水，诗梦暖于春”，落款“陈鸿寿”，下有一方形大印“渐入佳境”。

"Cold Pond Moon" Qin made by ZhaoWen, Banana Leaf Style.

Overall length 124 cm, effective chord length 111.6 cm, forehead width 18 cm, shoulder width 22 cm, tail width 15 cm, and thickness 5.3 cm.

The banana leaf-style "Cold Pond Moon" Qin, whose surface panel is made of Chinese fir, and bottom panel cypress, has a topcoat of red through black, with ebony accessories. Its cornu cervi degelatinatum lacquer cement can be partially seen. Inside the qin belly, it writes in ink "Handmade by ZhaoWen in the autumn of 2015 in Guangling, subscribed by A Yu of Mingtang". The name "Cold Pond Moon" on the bottom is inscribed by Mr. Ru Xiaoshi, a master of inscription in Xi'an. On both sides of Longchi, there is the line "The sound of the qin is as clear as water, the dreams of poetry are warmer than spring", signed by "Chen Hongshou" and a large square seal "As Good As It Gets".

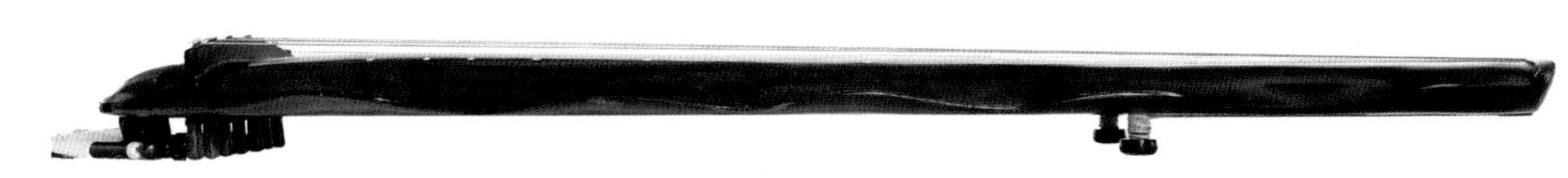

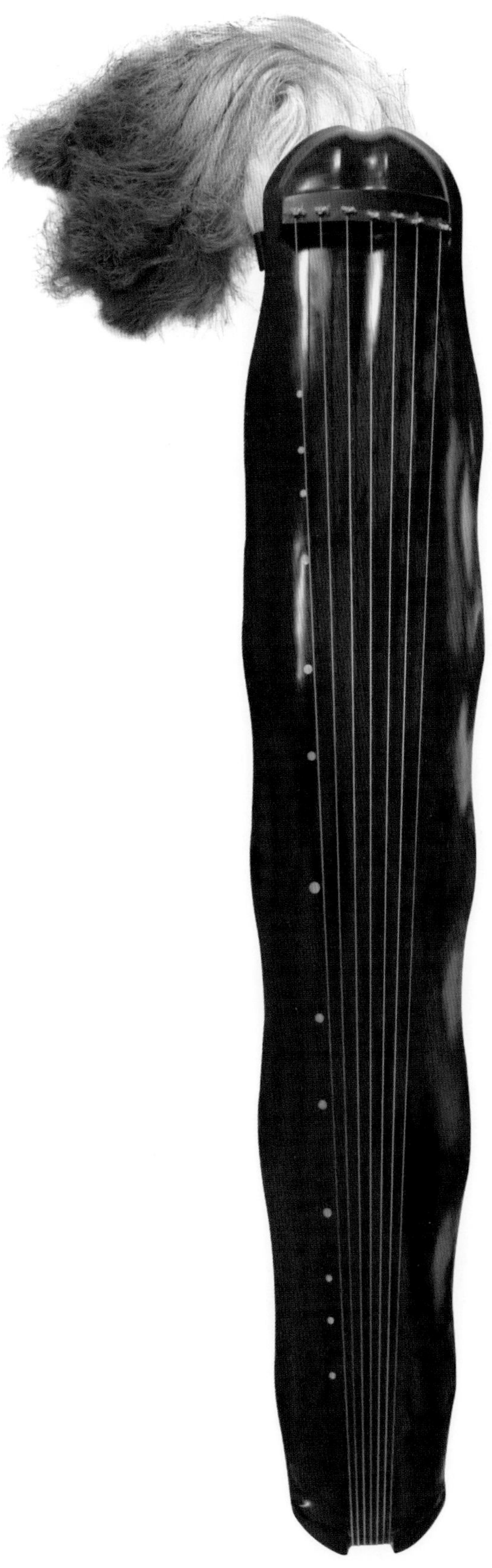

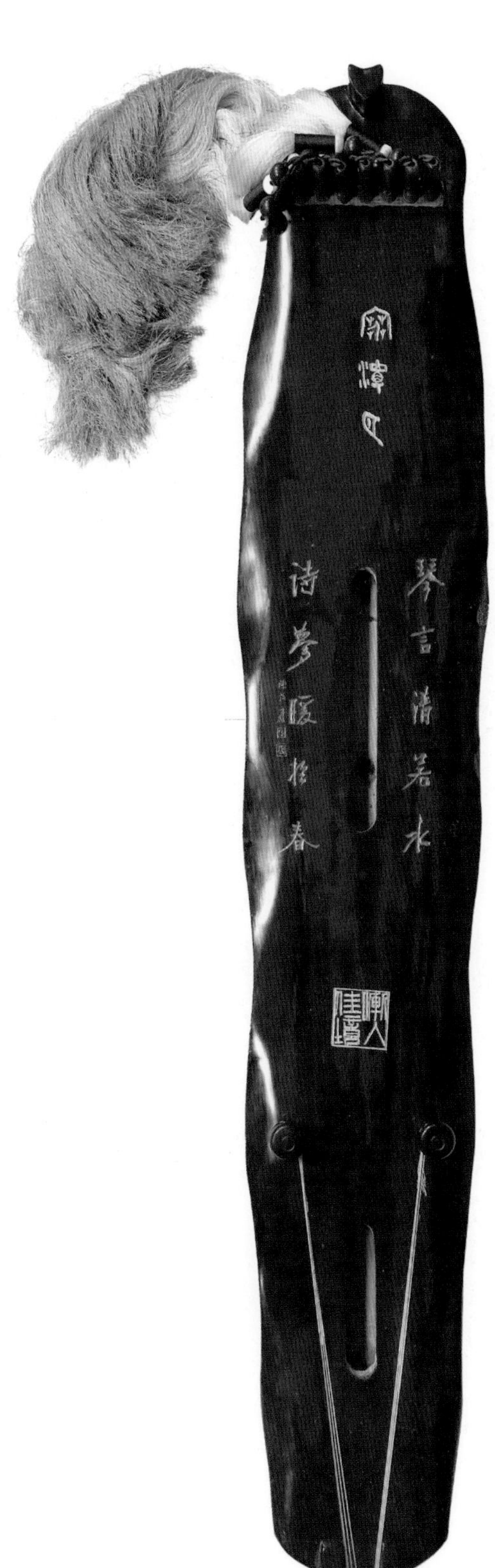
琴言清若水
诗梦暖於春

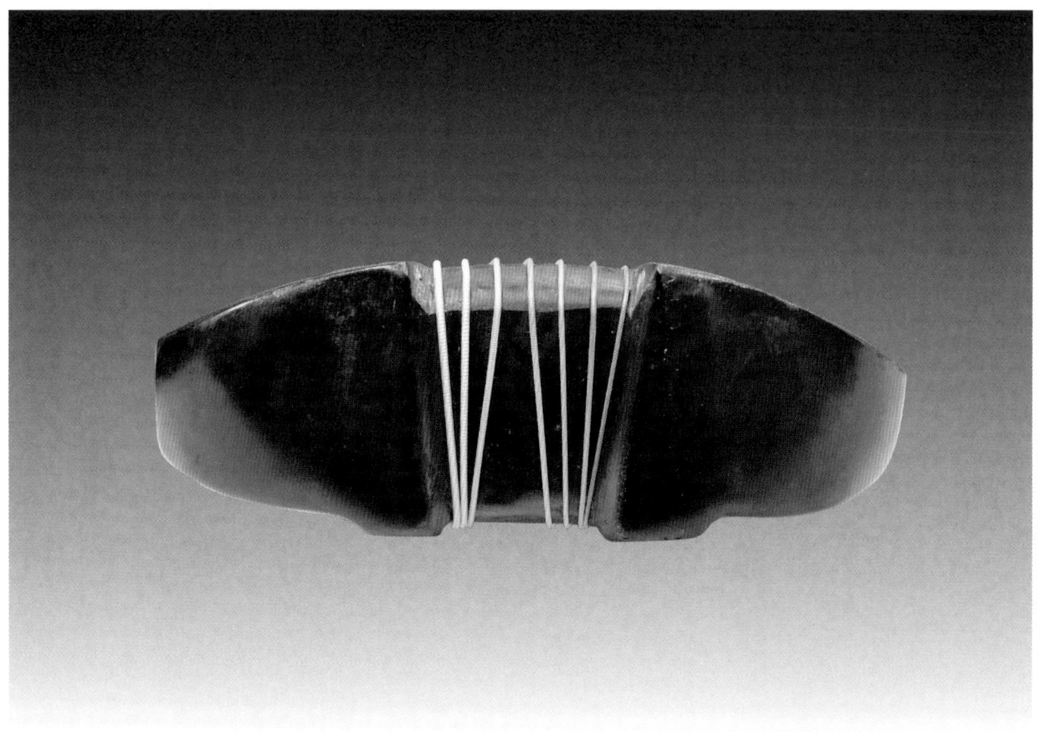

漸入佳境

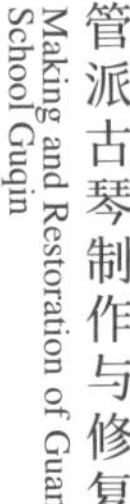

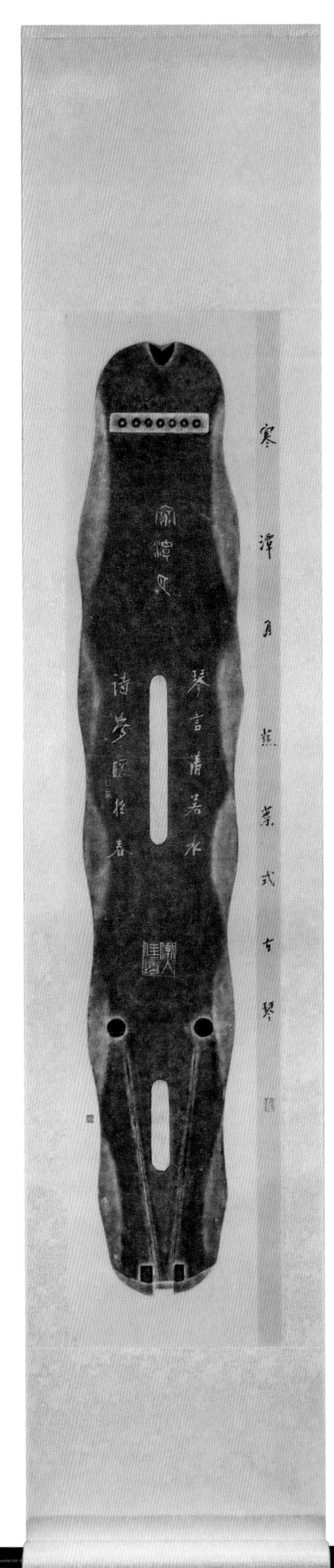
寒潭月蕉叶式古琴
琴言清若水

昭闻斫“连天碧”琴 仲尼式

通长 126 厘米，有效弦长 115 厘米，额宽 18 厘米，肩宽 20.3 厘米，尾宽 15 厘米，厚 5.6 厘米。

仲尼式“连天碧”琴，杉木面板，梓木底板，通体黑色，略透鹿角霜灰胎，乌木配件。琴腹内墨书“丙申荷月昭闻斫于长安，茗堂阿愚署”，琴底刻小篆名“连天碧”为西安文博金石大家茹小石先生题，龙池两侧刻有集句“汲古得修绠，开琴弄清弦”，落款“朱彝尊”。凤沼上有一长方形印章“陶然有余欢”。

"Boundless Green" Qin made by ZhaoWen, Zhong Ni Style.

Overall length 126 cm, effective chord length 115 cm, forehead width 18 cm, shoulder width 20.3 cm, tail width 15 cm, and thickness 5.6 cm.

The Zhong Ni-style "Boundless Green" Qin's surface panel is made of Chinese fur, and bottom plate catalpa. The whole body is covered by black paint, supplemented with ebony accessories. Inside the qin belly, it writes in ink "Made by ZhaoWen in the summer of 2016 in Chang'an, subscribed by A Yu of Mingtang". The seal name "Boundless Green" on the qin bottom is inscribed by Mr. Ru Xiaoshi, a master of inscription in Xi'an. On both sides of Longchi, there is the line "Perseverance is a must to draw from the ancients; only by dismantling a qin can we figure out how each string functions", with the inscription of "Zhu Yizun" and a rectangular seal "Tao Ran Yu Huan" above Fengzhao.

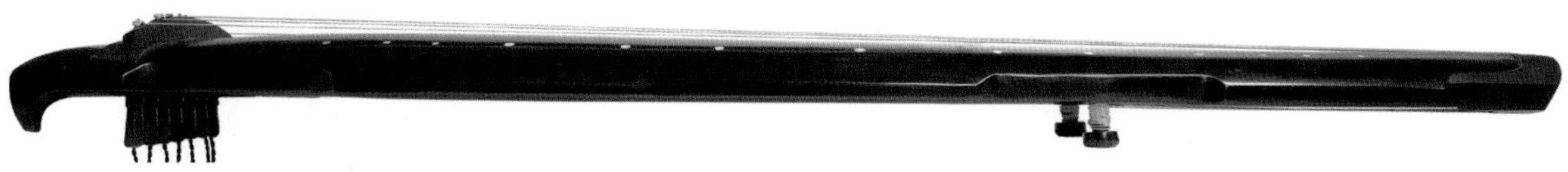

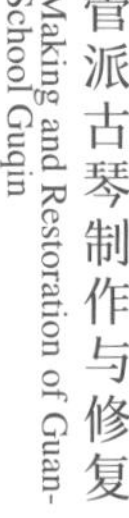

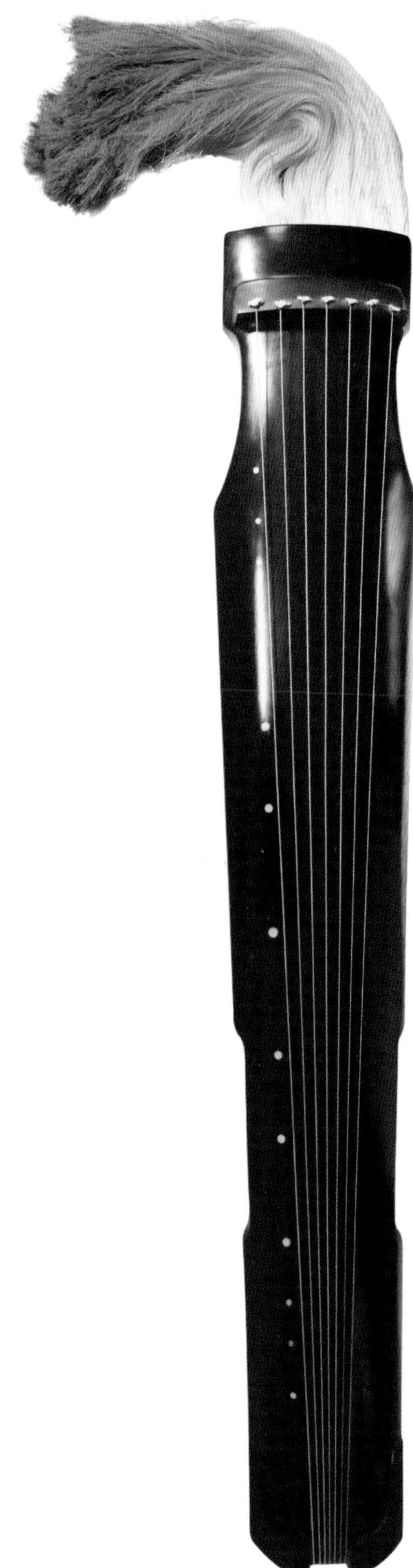

汲古得脩綆
開琴弄清緒

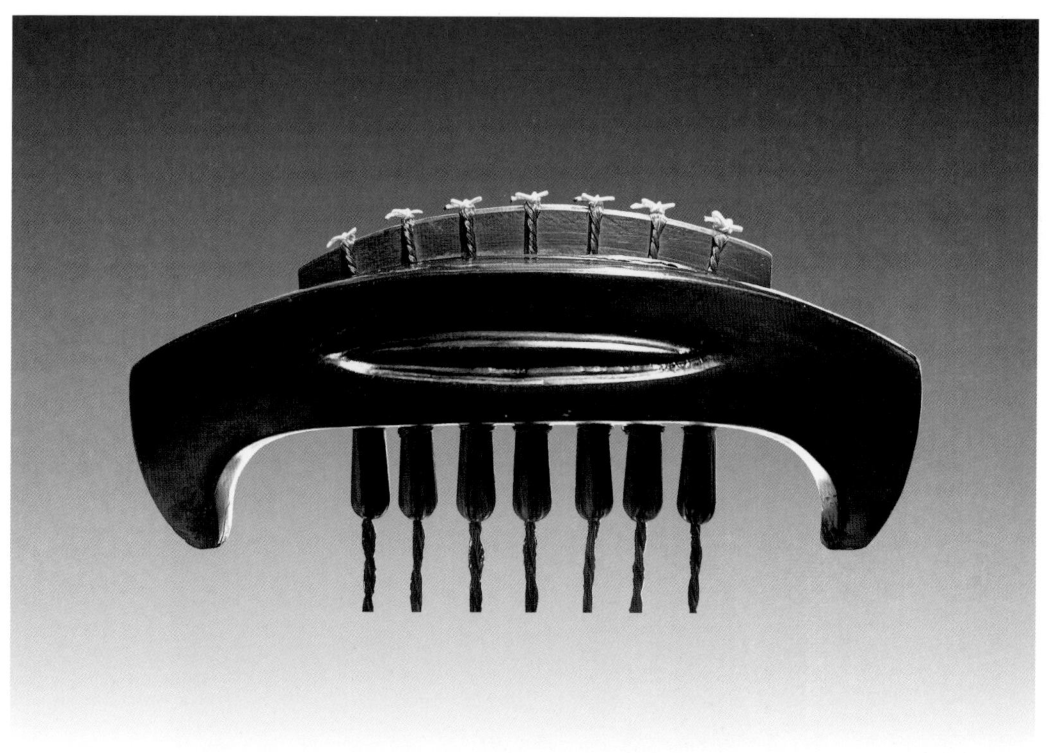

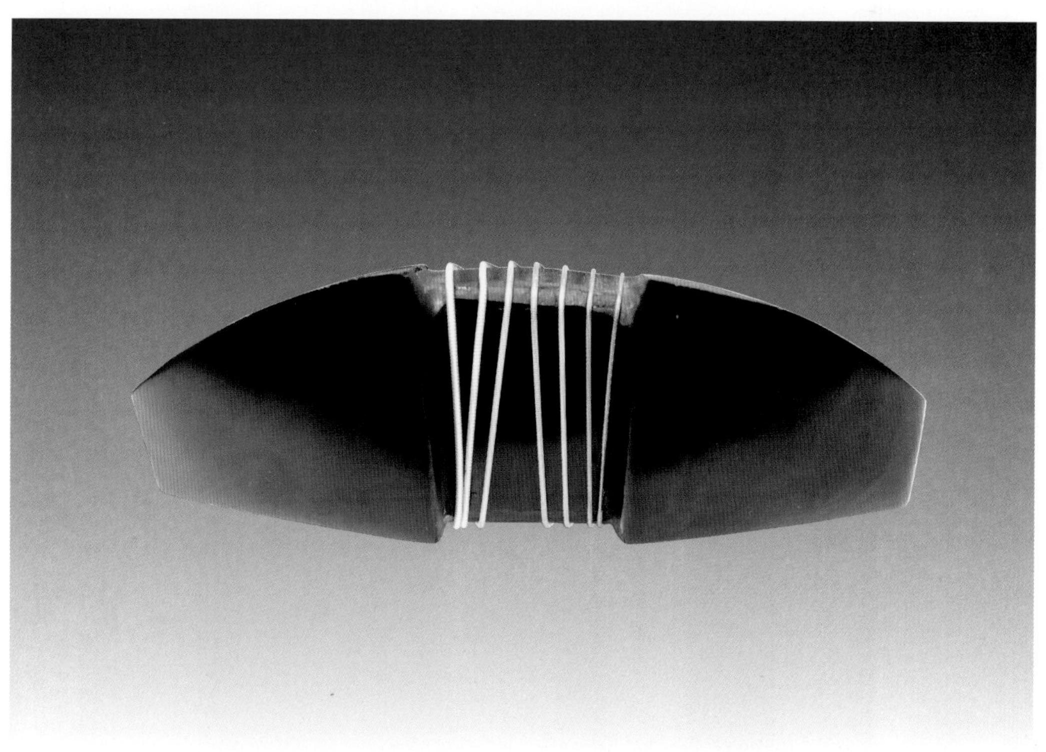

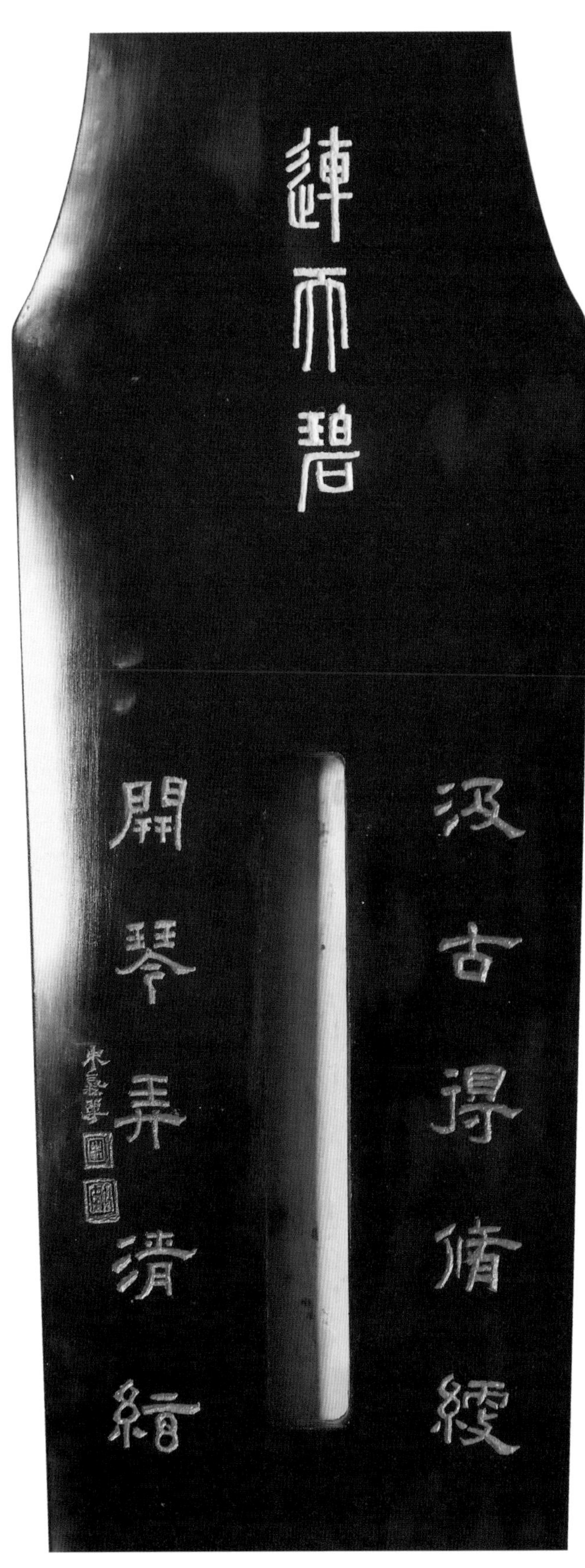
汲古得脩綆

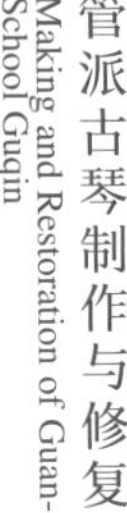

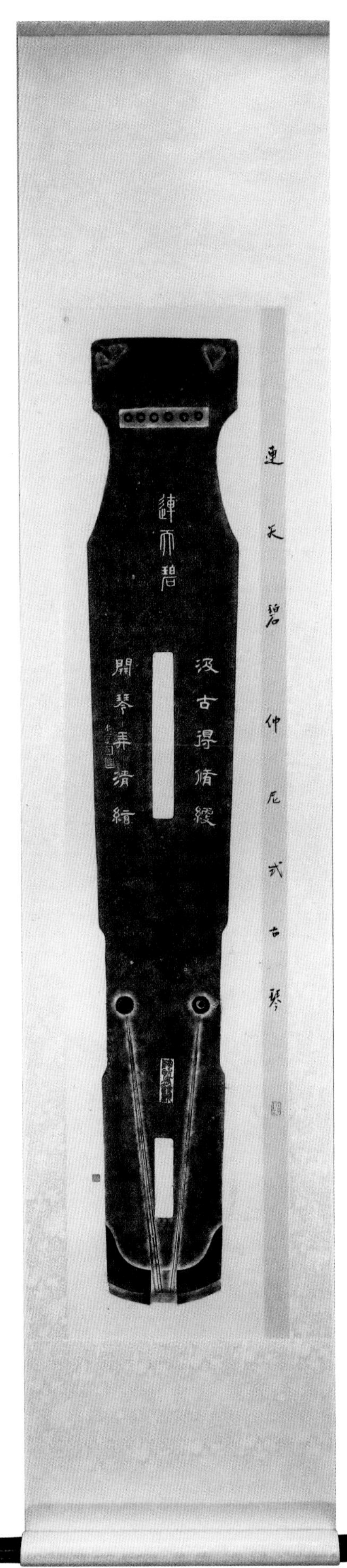
連天碧仲尼式古琴
連天碧
汲古得脩綆
開琴弄清絃

昭闻斫“燕掠暖溪”琴 净瓶式

通长120厘米，有效弦长117厘米，额宽19厘米，肩宽22厘米，尾宽14.8厘米，厚6.0厘米。

净瓶式“燕掠暖溪”琴，杉木面板，梓木底板，通体翠绿色，略透鹿角霜灰胎，黄花梨配件。琴腹内墨书“己亥年冬月昭闻重修于长安”，琴底刻名“燕掠暖溪”，左下方有“匋”字小印一枚，由西安文博金石大家茹小石先生题。下有韩春涛先生制“昭闻制琴”方印一枚，琴腰部有“澄怀”长方形印一枚。两雁足之间刻制有“琴”图形，为中国古琴博物馆标志。

"Swallow Skims over Creek" Qin made by ZhaoWen, Clean Bottle Style.

Overall length 120 cm, effective chord length 117 cm, forehead width 19 cm, shoulder width 22 cm, tail width 14.8 cm, and thickness 6.0 cm.

The clean bottle-style "Swallow Skims over Creek" Qin, whose surface panel is made of Chinese fur, and bottom panel catalpa, is covered by emerald topcoat, with the cornu cervi degelatinatum lacquer cement exposed slightly, and has fragrant rosewood accessories. Inside the qin belly, it writes in ink "Made by ZhaoWen in the winter of 2019 in Chang'an". The name "Swallow Skims over Creek" is inscribed on the bottom, and on its lower left, there is a small seal reading "Tao" by Mr. Ru Xiaoshi, a master of inscription in Xi'an, together with a square seal made by Mr. Han Chuntao, which reads "Qin Made by ZhaoWen". A rectangular seal reading "Cheng Huai" sits at the waist of the qin. The pattern " 琴 "is engraved between the two YanZu, which is the symbol of the Chinese Guqin Museum.

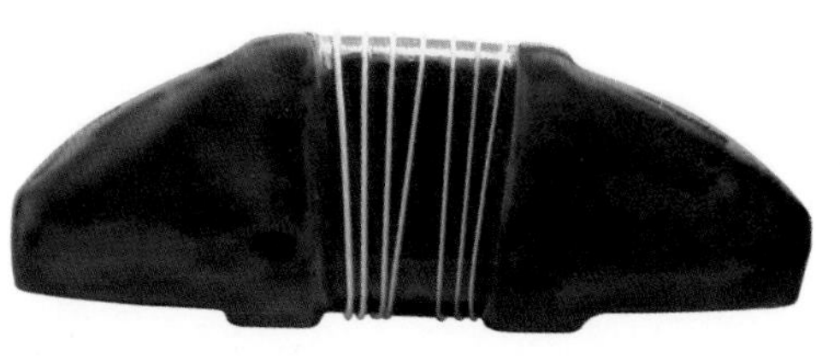

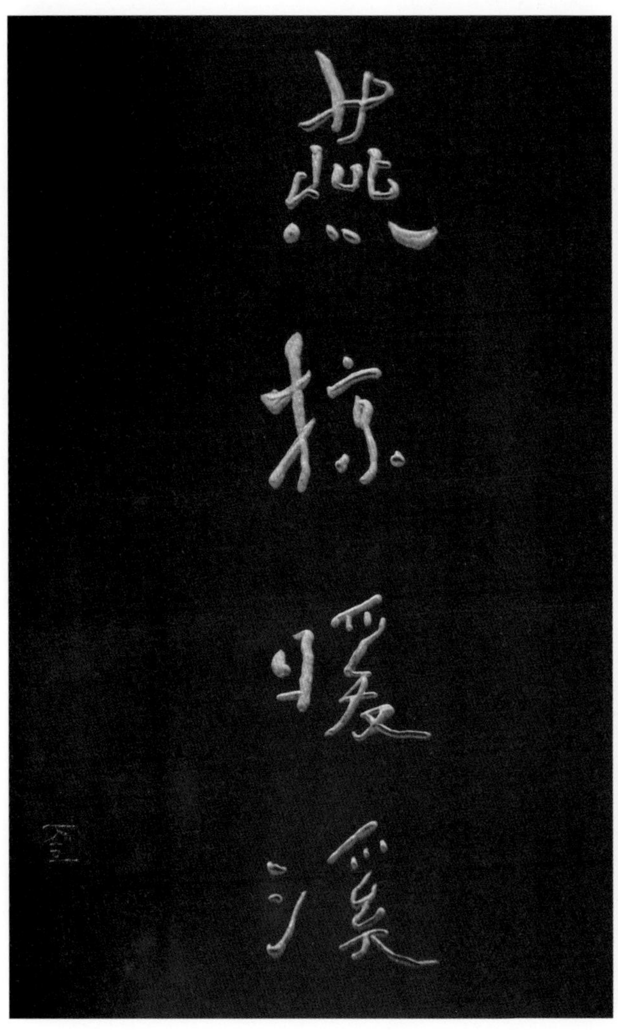

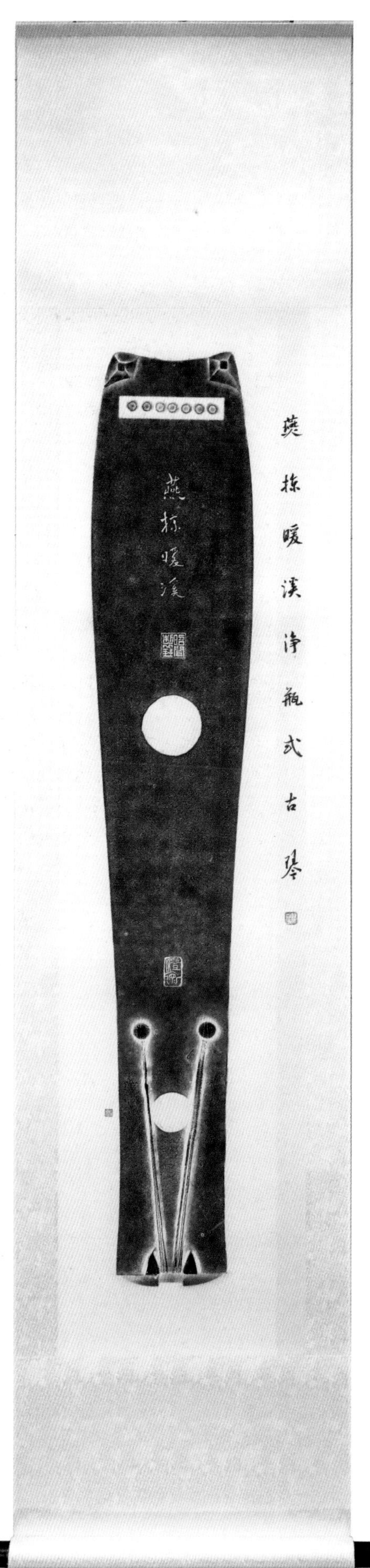
燕拣暖溪
燕拣暖溪净瓶式古琴

昭闻斫“烈日松风”琴 宣和式

通长 123.5 厘米，有效弦长 110.2 厘米，额宽 19 厘米，肩宽 19.8 厘米，尾宽 14.8 厘米，厚 5.8 厘米。

宣和式“烈日松风”琴，西安都城隍庙 660 年老松木房梁面板，香椿木底板，通体黑色，乌木配件。琴腹内墨书“丁酉之春昭闻斫于长安，茗堂阿愚署”，琴底刻名“烈日松风”，左下方有“匋丁”小印一枚，由西安文博金石大家茹小石先生题。下有韩春涛先生制“昭闻制琴”方印一枚。琴腰部有“凿池种莲”长方形印章一枚，是西安书法篆刻家钟镝先生制。

"Burning Sun-Pine-Wind" Qin made by ZhaoWen, Xuanhe Style.

Overall length 123.5 cm, effective chord length 110.2 cm, forehead width 19 cm, shoulder width 19.8 cm, tail width 14.8 cm, and thickness 5.8 cm.

The Xuanhe-style "Burning Sun-Pine-Wind" Qin, whose surface panel is made of a 660-year-old pine beam from the Capital City God Temple, and bottom panel Toona sinensis wood, has black topcoat and ebony accessories. Inside the qin belly, it writes in ink "Made by ZhaoWen in the spring of 2017 in Chang'an, subscribed by A Yu of Mingtang". The name "Burning Sun-Pine-Wind" is inscribed on the bottom, and on its lower left, there is a small seal reading "Tao Ding" by Mr. Ru Xiaoshi, a master of inscription in Xi'an, together with a square seal made by Mr. Han Chuntao, which reads "Qin Made by ZhaoWen". A rectangular seal reading "Chisel Pond to Plant Lotus" sits at the waist of the qin, produced by Mr. Zhong Di, a master of calligraphy and seal engraving in Xi'an.

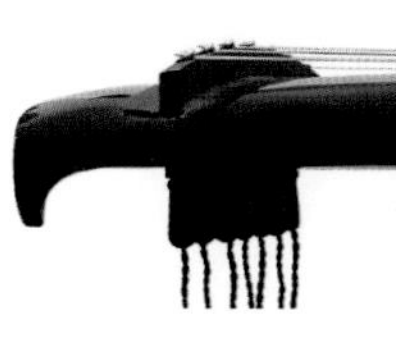

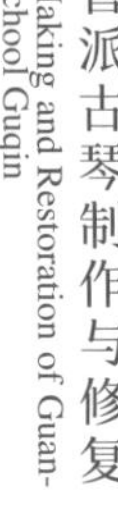

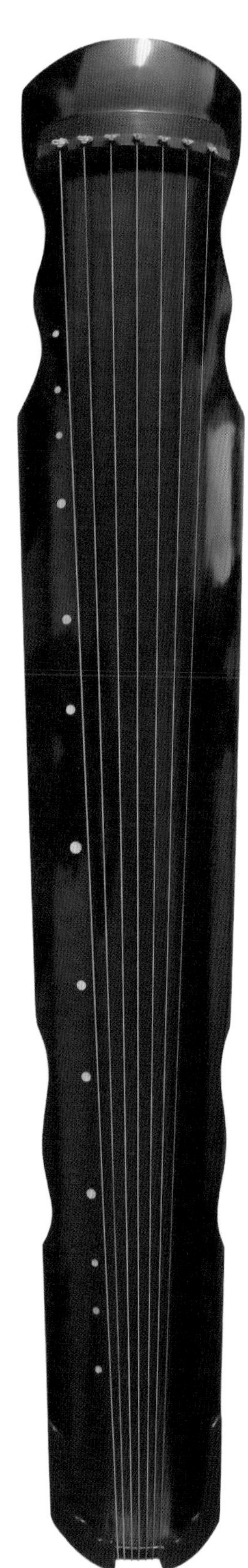

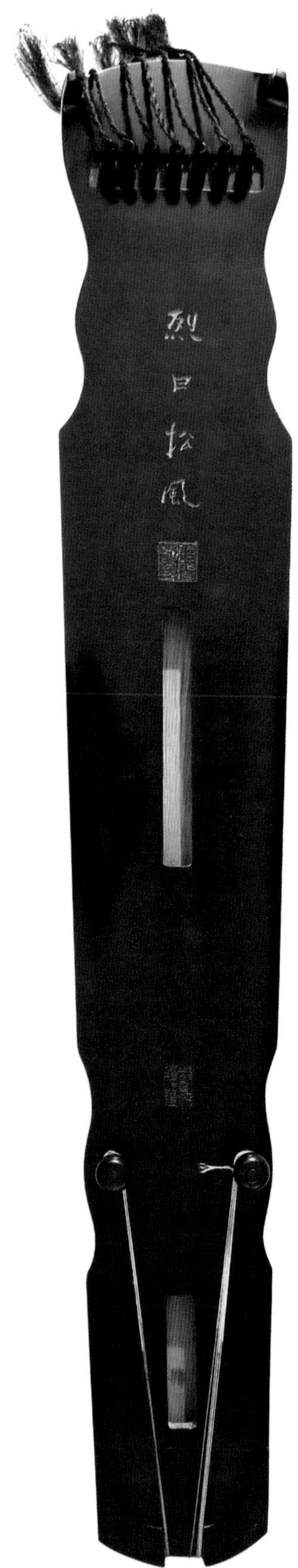
烈日松風

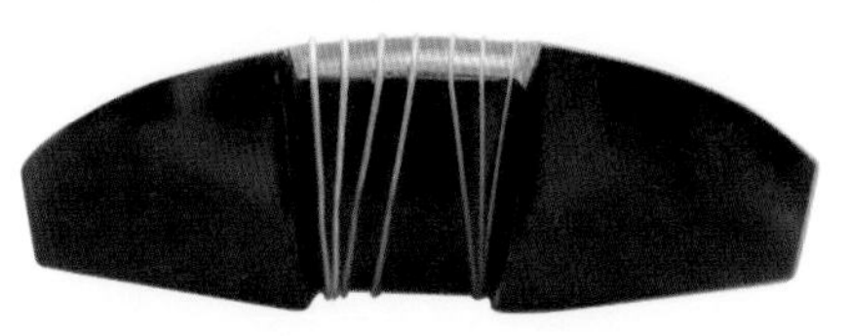

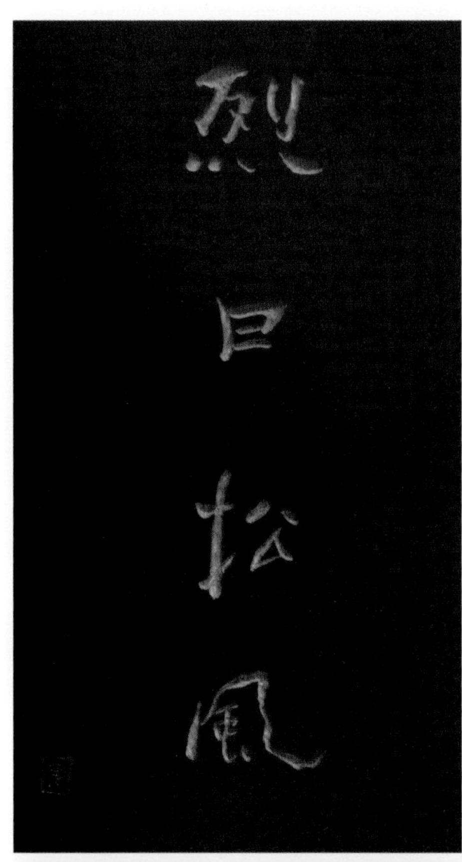
烈日松風

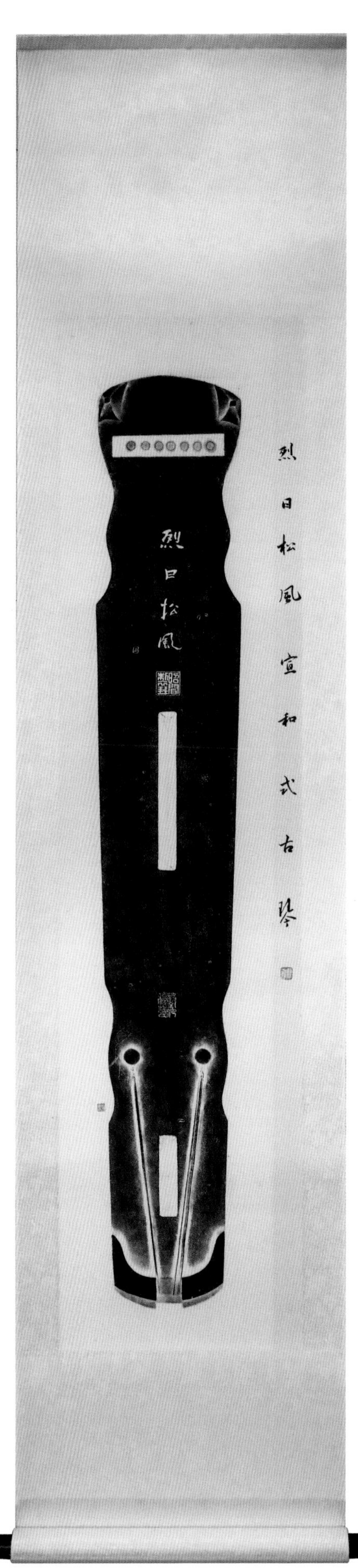
烈日松风
烈日松风宣和式古琴

昭闻斫“霜折焦枝”琴 神农式

通长121厘米，有效弦长109.6厘米，额宽20厘米，肩宽20.8厘米，尾宽14.0厘米，厚5.6厘米。

神农式“霜折焦枝”琴，西安都城隍庙660年老松木房梁面板，香椿木底板，通体黑色，乌木配件。琴腹内墨书“乙未年昭闻斫于长安，茗堂阿愚署”，琴底刻名“霜折焦枝”，左下方有“陶丁金石”小印一枚，由西安文博金石大家茹小石先生题。下有韩春涛先生制“昭闻制琴”方印一枚。琴腰部有“大方无隅”方形印章一枚。

"Banana Leaves Frozen by Killing Frost" Qin made by ZhaoWen, Shen Nong Style.

Overall length 121 cm, effective chord length 109.6 cm, forehead width 20 cm, shoulder width 20.8 cm, tail width 14.0 cm, and thickness 5.6 cm.

The Shen Nong Style "Banana Leaves Frozen by Killing Frost" Qin, whose surface panel is made of a 660-year-old pine beam from the Capital City God Temple, and bottom panel Toona sinensis wood, has black topcoat and ebony accessories. Inside the qin belly, it writes in ink "Made by ZhaoWen in 2015 in Chang'an, subscribed by A Yu of Mingtang". The name "Banana Leaves Frozen by Killing Frost" is inscribed on the bottom, and on its lower left, there is a small seal reading "Tao Ding Jin Shi" by Mr. Ru Xiaoshi, a master of inscription in Xi'an, together with a square seal made by Mr. Han Chuntao, which reads "Qin Made by ZhaoWen". A square seal reading "The Largest Square Has No Corners" sits at the waist of the qin.

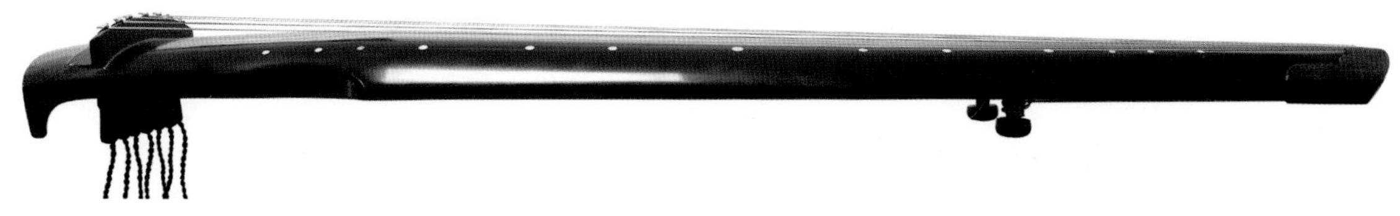

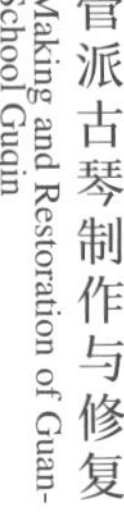

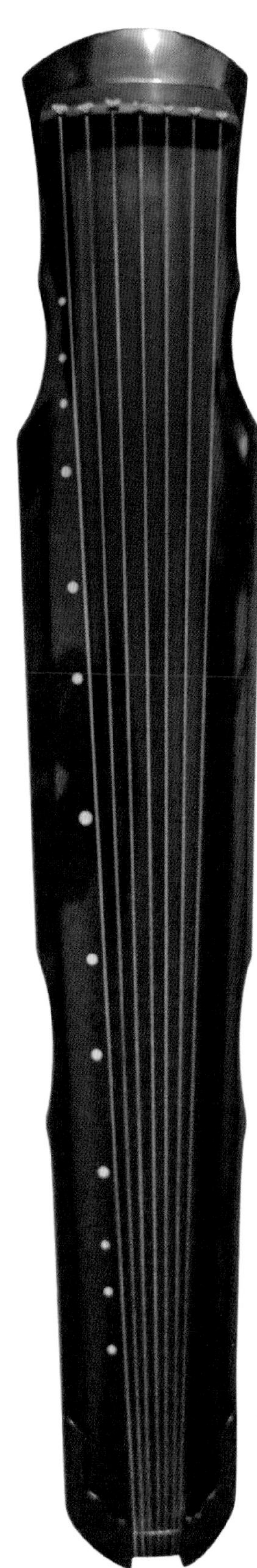

霜折焦枝

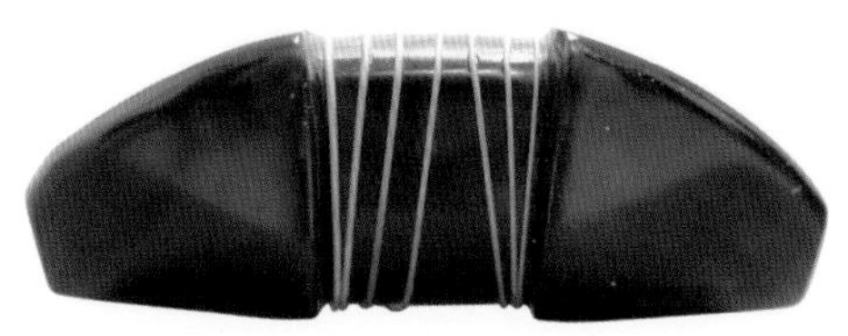

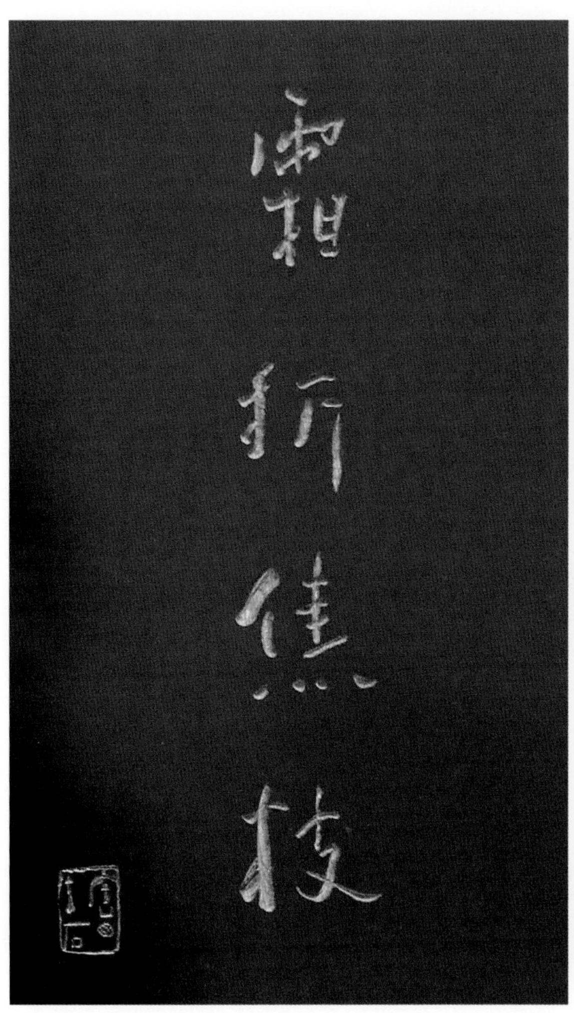

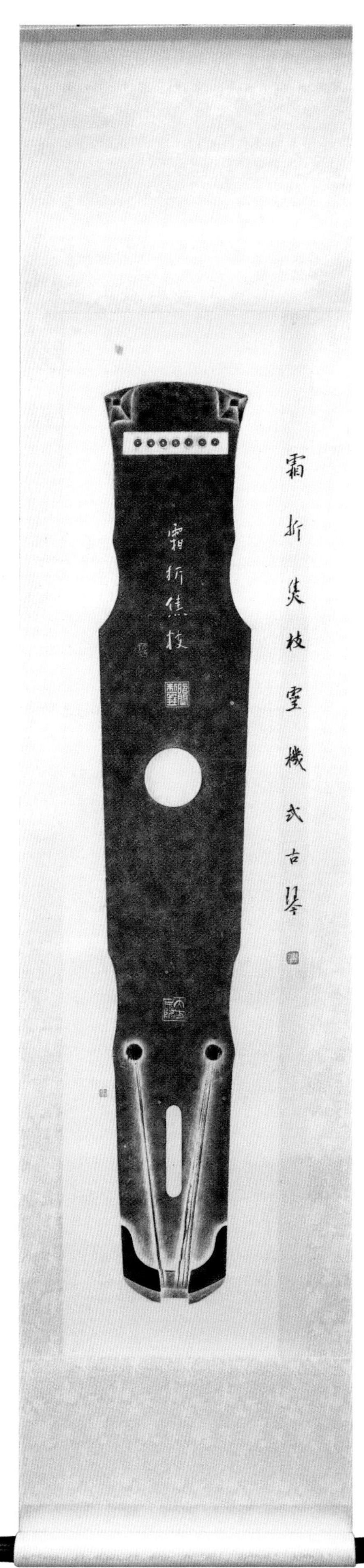
霜折焦枝
霜折焦枝靈機式古琴

昭闻斫“落雪卷沙”琴 仲尼式

通长 122 厘米，有效弦长 109.7 厘米，额宽 18 厘米，肩宽 20.0 厘米，尾宽 13.8 厘米，厚 5.8 厘米。

仲尼式“落雪卷沙”琴，老杉木房梁面板，楸木底板，通体栗壳色，黑檀配件。琴腹内墨书“甲午年昭闻斫于长安”，琴底刻名“霜落雪卷沙”，左下方有“丁”字小印一枚，由西安文博金石大家茹小石先生题。下有韩春涛先生制“昭闻制琴”方印一枚。琴腰部有“坐观众妙”方形印章一枚。

"Falling Snow and Rolling Sand" Qin made by ZhaoWen, Zhong Ni Style.

Overall length 122 cm, effective chord length 109.7 cm, forehead width 18 cm, shoulder width 20.0 cm, tail width 13.8 cm, and thickness 5.8 cm.

The Zhong Ni-style "Falling Snow and Rolling Sand" Qin, whose surface panel is made of Chinese fir beam, and bottom Catalpa wood, has chestnut shell topcoat and ebony accessories. Inside the qin belly, it writes in ink "Made by ZhaoWen in 2014 in Chang'an". The name "Falling Snow and Rolling Sand" is inscribed on the bottom, and on its lower left, there is a small seal "Ding" by Mr. Ru Xiaoshi, a master of inscription in Xi'an, together with a square seal made by Mr. Han Chuntao, which reads "Qin Made by ZhaoWen". A square seal reading "Sit Down and See the World" sits at the waist of the qin.

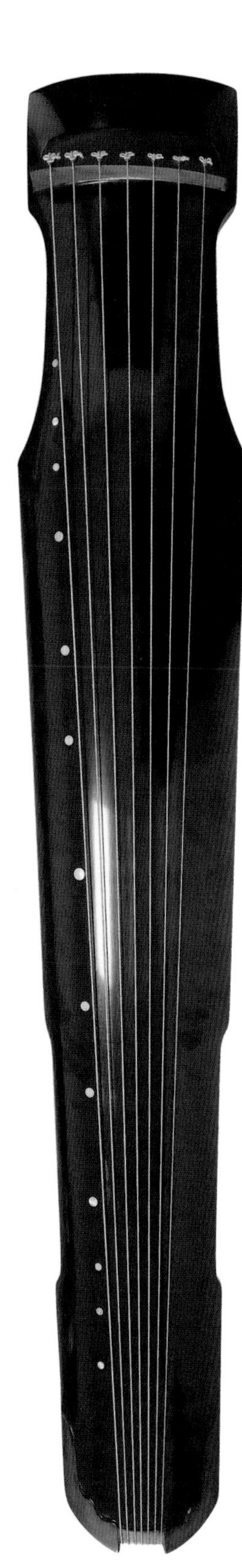

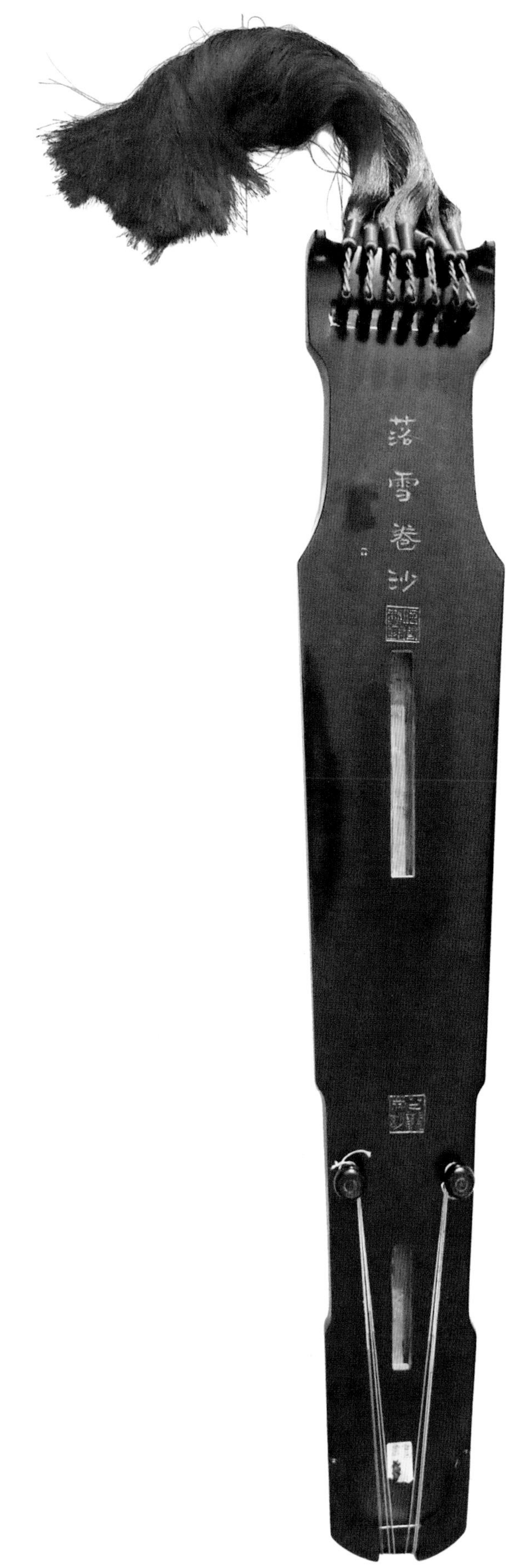
落雪卷沙

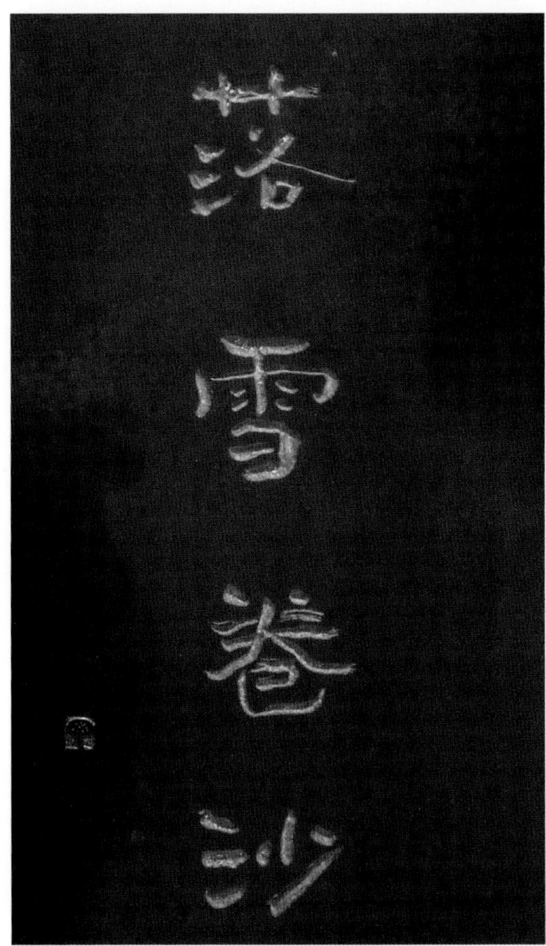
落雪卷沙

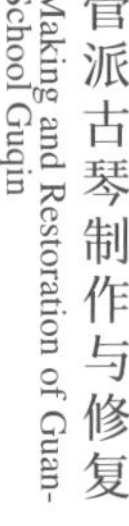

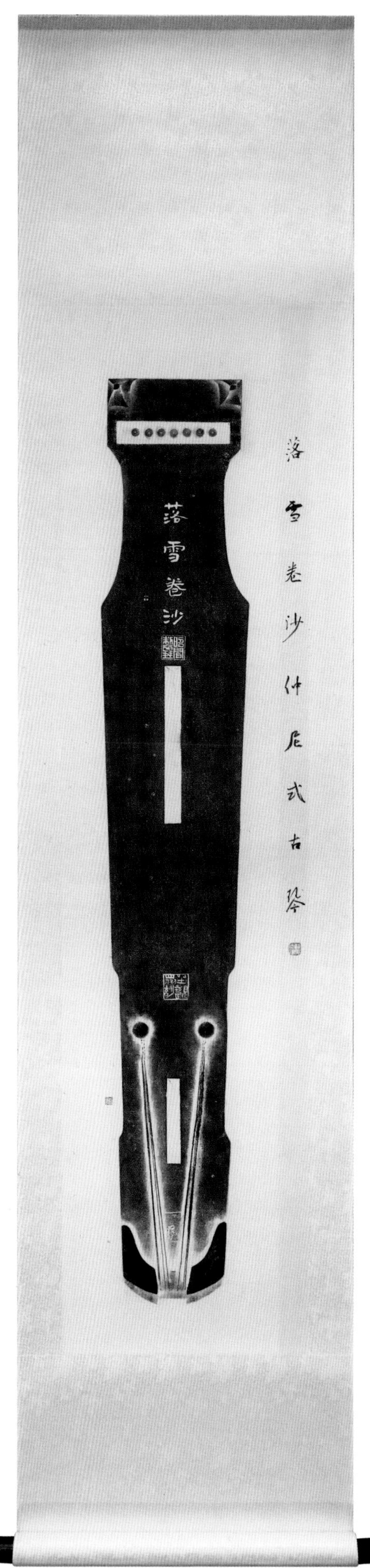
落雪卷沙
落雪卷沙仲尼式古琴

昭闻斫“忘机”琴 神农式

通长 121 厘米，有效弦长 109.5cm 额宽 19.2 厘米，肩宽 20.0 厘米，尾宽 13.5 厘米，厚 5.6 厘米。

神农式“忘机”琴，西安都城隍庙 660 年老松木房梁面板，香椿木底板，通体黑色，乌木配件。琴腹内墨书“乙未秋昭闻斫于长安，茗堂阿愚署”，琴底刻名“忘机”，左侧“辛丑二月春涛书”，下面“昭闻制琴”方印，龙池两侧“莫道无弦真有趣，须于弦上悟无弦”均为西安书法篆刻家韩春涛先生题制。琴腰部有“似花非花”长方形印章一枚。

"Standing Aloof from the World" Qin made by ZhaoWen, Shen Nong Style.

Overall length 121 cm, effective chord length 109.5 cm, forehead width 19.2 cm, shoulder width 20.0 cm, tail width 13.5 cm, and thickness 5.6 cm.

The Shen Nong Style "Standing Aloof from the World" Qin, whose surface panel is made of a 660-year-old pine beam from the Capital City God Temple, and bottom panel Toona sinensis wood, has black topcoat and ebony accessories. Inside the qin belly, it writes in ink "Made by ZhaoWen in the autumn of 2015 in Chang'an, subscribed by A Yu of Mingtang". The bottom is inscribed with the name "Standing Aloof from the World" and the line "Wrote by Chuntao on February of 2021" on its left, together with a square seal that reads "Qin Made by ZhaoWen" below. The lines "Never say that fun lies not in the strings, but to realize the state of no string through the strings" on both sides of the Longchi are engraved by Mr. Han Chuntao, a master of calligraphy and seal engraving in Xi'an. On the waist of the qin, there is a rectangular seal reading "Flowers Yet Not Flowers".

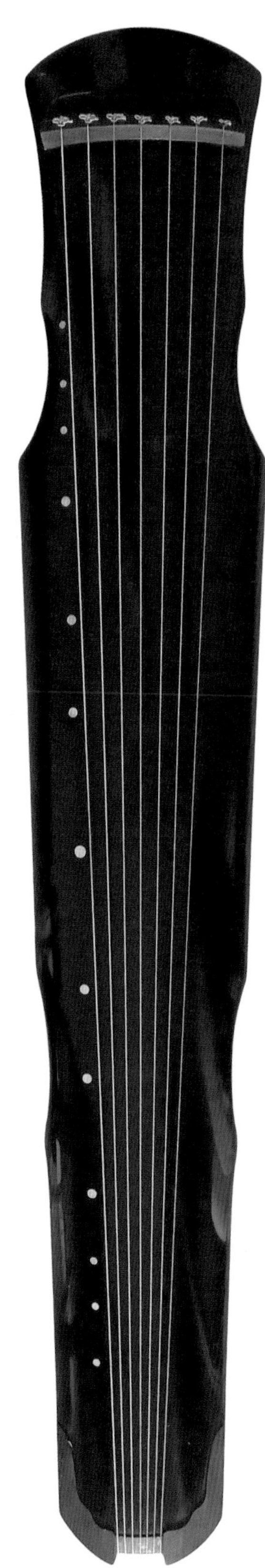

莫道无弦真有趣
须於弦上悟无弦

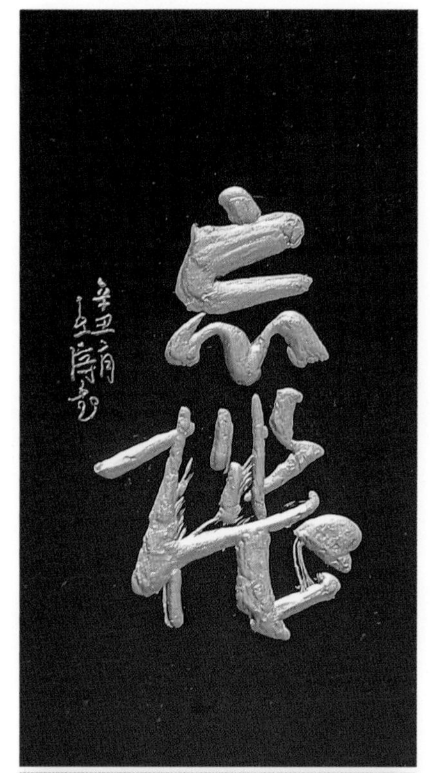

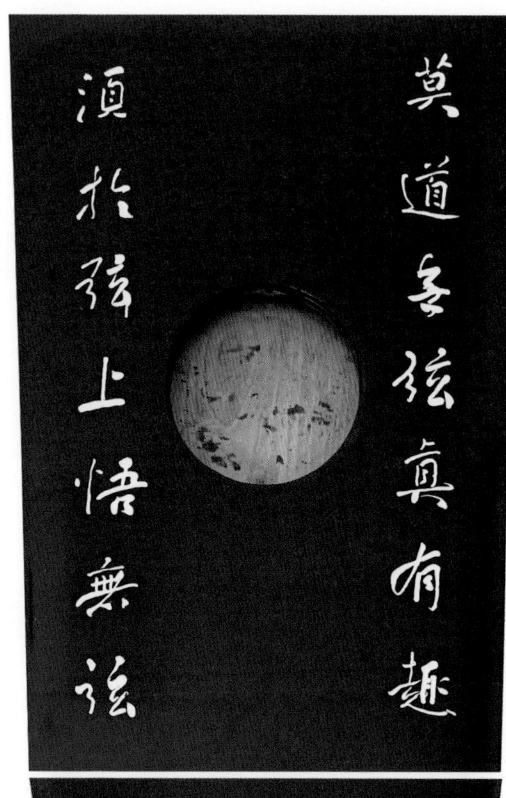
莫道無弦真有趣
須於弦上悟無弦

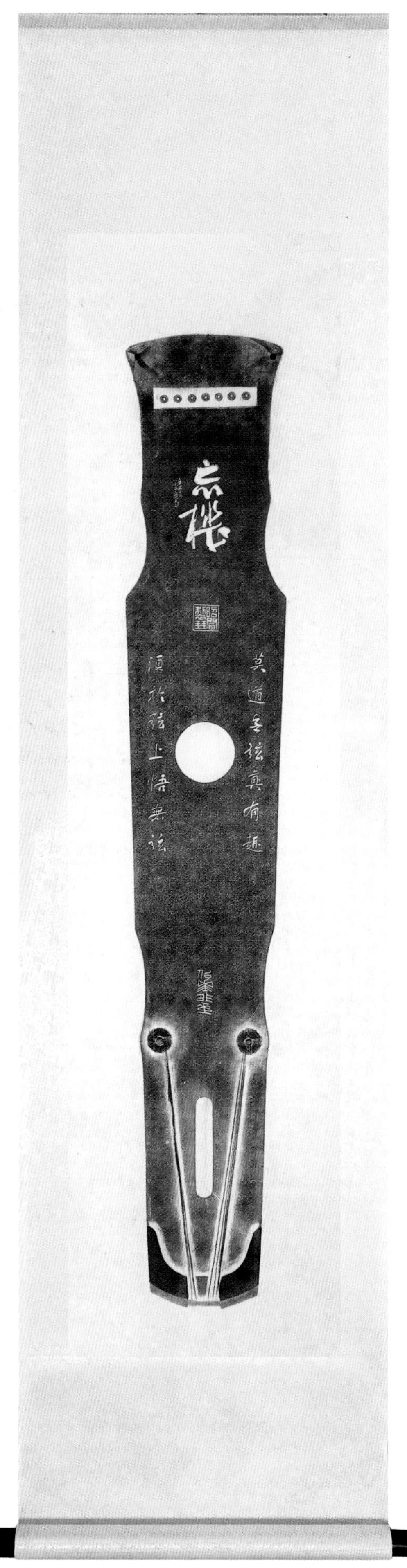
莫道無弦真有趣
須於弦上悟無弦

昭闻斫“松风”琴 仲尼式

通长 126 厘米，有效弦长 115.8 厘米，额宽 18 厘米，肩宽 20.3 厘米，尾宽 14.5 厘米，厚 5.5 厘米。

仲尼式“松风”琴，西安都城隍庙 660 年老松木房梁面板，香椿木底板，通体黑色透金，乌木配件。琴腹内墨书“乙未年秋月昭闻斫于长安，茗堂阿愚署”，琴底刻名“松风”，左有“质文”印一枚，下有韩春涛先生制“昭闻制琴”印一枚。龙池两侧的“有花有酒春常在，多情多兴寿自高”由西安美术学院陈斌教授题写。琴腰部有韩春涛制“琴酒伴溪云”印一枚。

"Pine-Wind" Qin made by ZhaoWen, Zhong Ni style.

Overall length 126 cm, effective chord length 115.8 cm, forehead width 18 cm, shoulder width 20.3 cm, tail width 14.5 cm, and thickness 5.5 cm.

The Zhong Ni-style "Pine-Wind" Qin, whose surface panel is made of a 660-year-old pine beam from the Capital City God Temple, and bottom panel Toona sinensis wood, has black and gold topcoat and ebony accessories. Inside the qin belly, it writes in ink "Made by ZhaoWen in the autumn of 2015 in Chang'an, subscribed by A Yu of Mingtang". The name "Pine-Wind" is inscribed on the bottom, and on its left, there is a small seal reading "Zhi Wen", together with a square seal made by Mr. Han Chuntao, which reads "Qin Made by ZhaoWen" below. The lines "Where there are flowers and wine, there is spring; where there are passion and joy, there is longevity" on both sides of the Longchi are inscribed by Prof. Chenbin of Xi'an Academy of Fine Arts. A seal reading "Playing Guqin and Drinking Wine Besides the Creek" made by Mr. Han Chuntao sits at the waist of the qin.

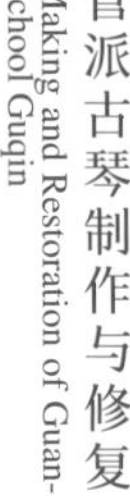

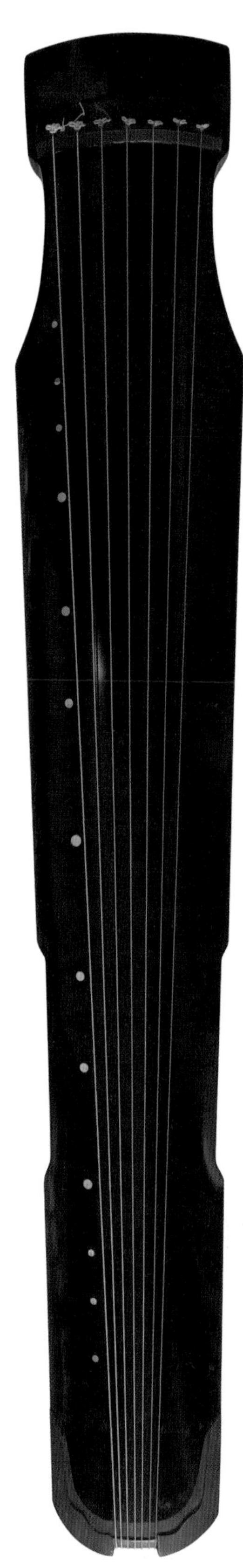

松風

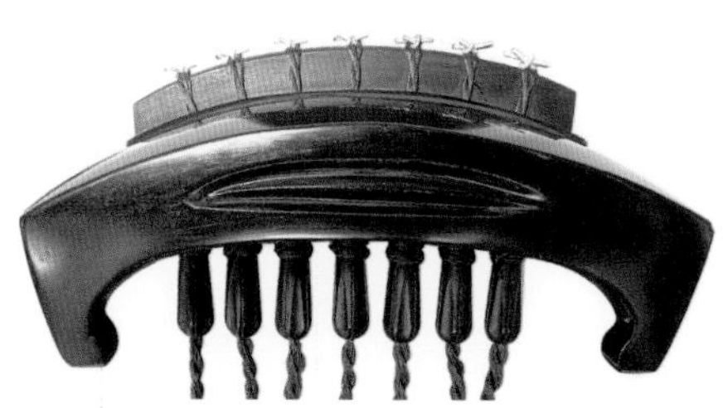

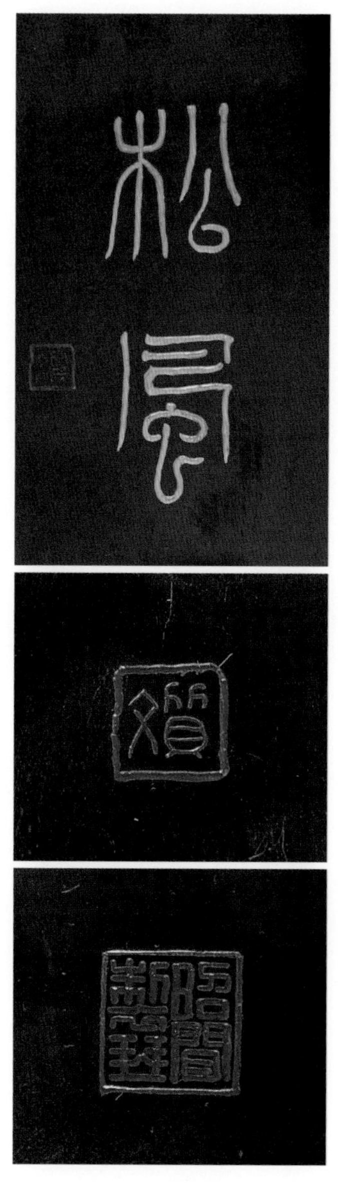

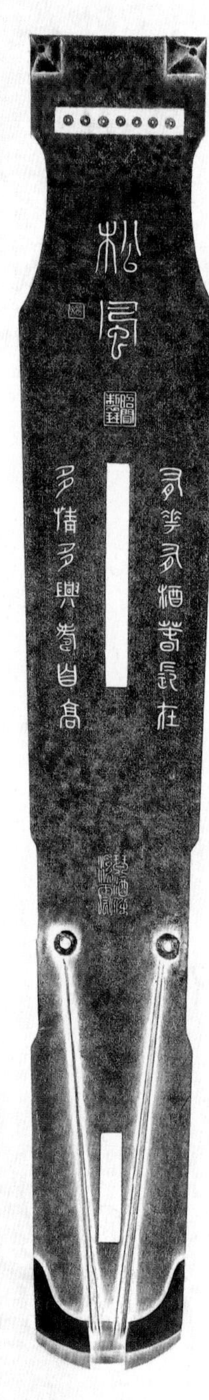
松風

昭闻斫“璇曧”琴 仲尼式

通长126厘米，有效弦长115.8厘米，额宽18厘米，肩宽20.3厘米，尾宽15厘米，厚5.5厘米。

仲尼式“璇曧”琴，杉木面板，楸木底板，通体黑色透金，乌木配件。琴腹内墨书“丙申荷月昭闻斫于长安，茗堂阿愚署”，琴底刻名“璇曧”由西安文博金石大家茹小石先生题，下有韩春涛先生制“昭闻制琴”印一枚。龙池两侧的“琴伴庭前月，衣无世外尘”由吴江文联主席孙俊良先生题写。琴腰部有韩春涛制“缥香”印一枚。

"Xuan Yuan" Qin made by ZhaoWen, Zhong Ni Style.

Overall length 126 cm, effective chord length 115.8 cm, forehead width 18cm, shoulder width 20.3 cm, tail width 15 cm, and thickness 5.5 cm.

The Zhong Ni-style "Xuan Yuan" Qin, whose surface panel is made of Chinese fir beam, and bottom Catalpa wood, has black and gold shell topcoat and ebony accessories. Inside the qin belly, it writes in ink "Made by ZhaoWen in the summer of 2016 in Chang'an, subscribed by A Yu of Mingtang". The seal name "Xuan Yuan" on the bottom of the qin is inscribed by Mr. Ru Xiaoshi, a master of inscription in Xi'an. Under it, is a square seal that reads "Qin Made by ZhaoWen", made by Mr. Han Chuntao. The lines "Playing the qin with the moonlight seems to have jumped out of the world of mortals." on both sides of the Longchi are by Chairman Sun Junliang of the Wujiang Federation of Literary and Art Circles. A seal reading "Ethereal Fragrance" sits at the waist of the qin.

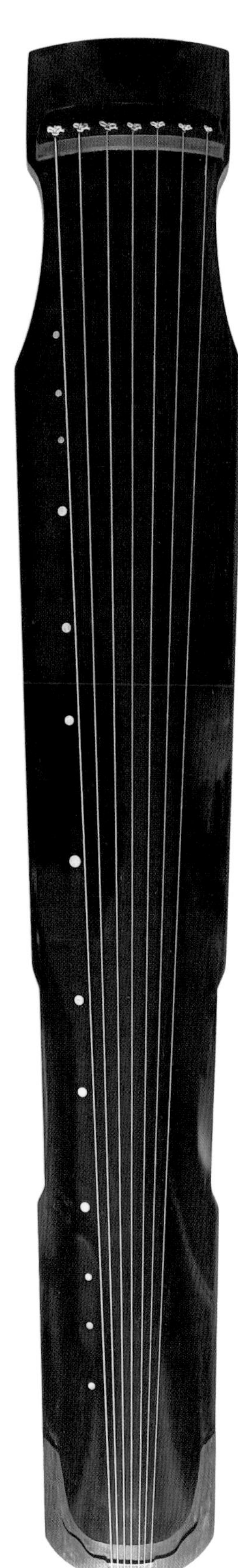

琴伴庭前月
衣無世外塵

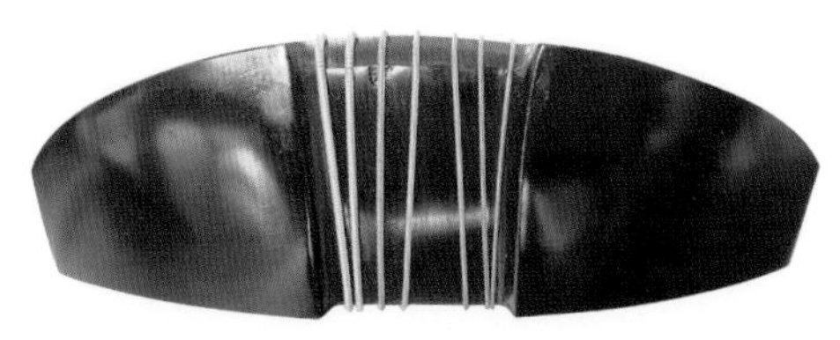

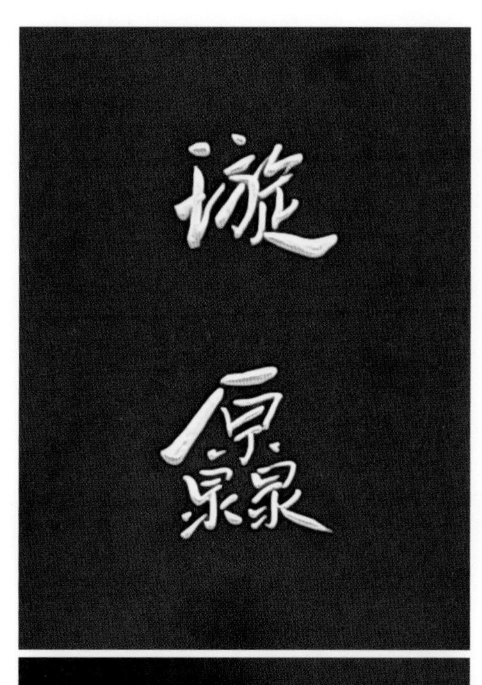

琴伴庭前月
衣無世外塵

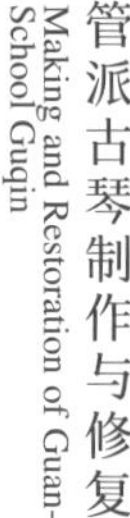

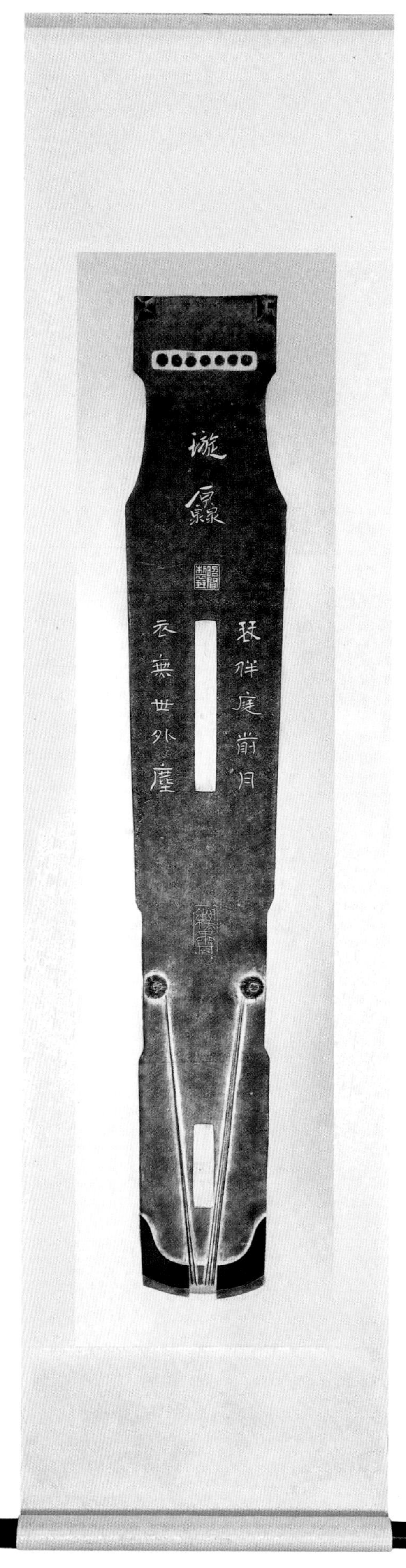
琴伴庭前月
衣無世外塵

昭闻斫“独幽”琴 混沌式

通长 126.0 厘米，有效弦长 113.2 厘米，额宽 17.5 厘米，肩宽 20.0 厘米，尾宽 15.0 厘米，厚 6.0 厘米。

混沌式“独幽”琴，西安都城隍庙 660 年老松木房梁面板，香椿木底板，通体黑漆，略透鹿角霜灰胎，乌木配件，龙池内可见灰胎修补古木旧痕。琴腹内墨书“乙未秋昭闻斫于长安，茗堂阿愚署”，琴底刻名“独幽”，左侧印章为“钟镝印信”，下有韩春涛先生制“昭闻制琴”印一枚。龙池两侧对联“看花听竹心无事，品茗焚香乐有余”由西安书法篆刻家钟镝先生题写。琴腰部有钟镝制“无为”印一枚。

"Alone in the Gloom" Qin made by ZhaoWen, Taichi-chaos Style.

Overall length 126.0 cm, effective chord length 113.2 cm, forehead width 17.5 cm, shoulder width 20.0 cm, tail width 15.0 cm, and thickness 6.0 cm.

The Taichi-chaos style "Alone in the Gloom" Qin's surface panel is made of a 660-year-old pine beam from the Capital City God Temple, and its bottom panel is made of Toona sinensis wood. The finishing coat is mainly black, blended with cornu cervi degelatinatum. It has ebony accessories, and old traces of repairment can be seen in its Longchi part. Inside the qin belly, it writes in ink "Made by ZhaoWen in the autumn of 2015 in Chang'an, subscribed by A Yu of Mingtang". The bottom is inscribed with the name "Alone in the Gloom" and the "Zhong Di's Seal" on its left, together with a seal that reads "Qin Made by ZhaoWen" made by Mr. Han Chuntao below. The lines "Inner peace can be gained by watching the flowers and listening to the sounds of bamboo, and there's so much fun in sipping tea and burning incenses" on both sides of the Longchi are by Mr. Zhong Di, a master of calligraphy and seal engraving in Xi'an. At the waist of the qin, there is the seal "Non-action" made by Zhong Di.

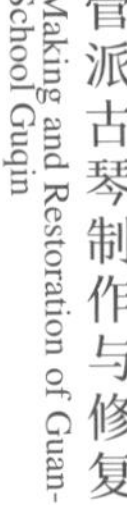

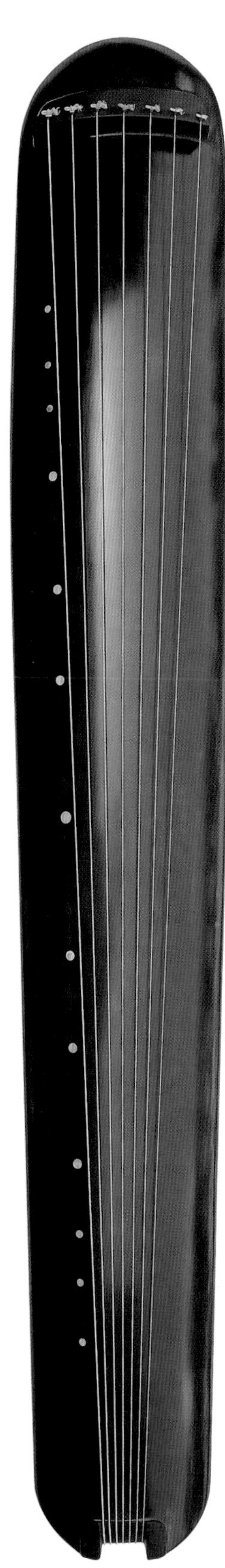

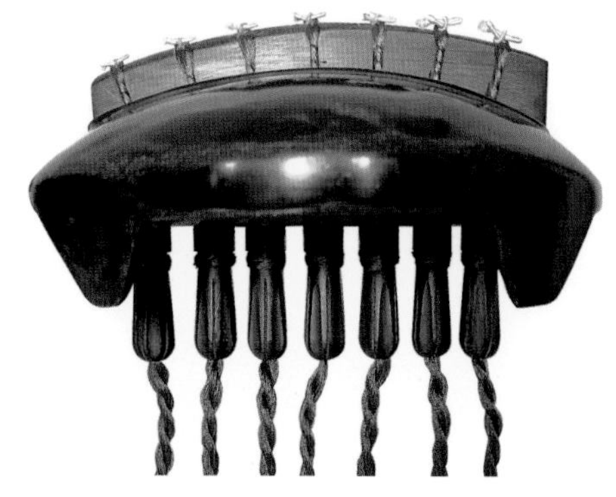

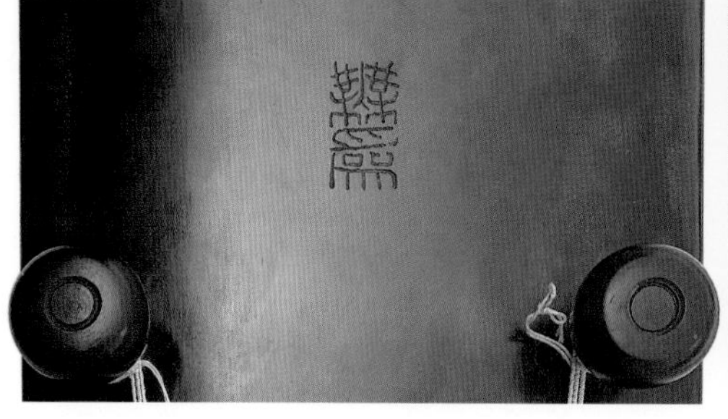

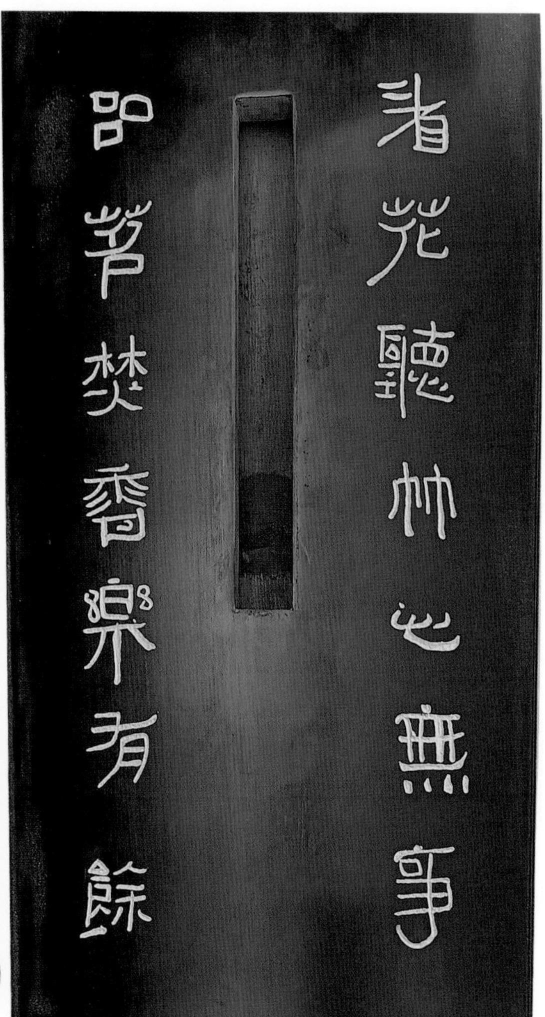
看花聽竹心無事
品茗焚香樂有餘

看花聽竹心無事
品茗焚香樂有餘

昭闻斫“蕉林听雨”琴 蕉叶式

通长 120.0 厘米，有效弦长 109.5 厘米，额宽 18.0 厘米，肩宽 19.5 厘米，尾宽 14.5 厘米，厚 5.0 厘米。

蕉叶式“蕉林听雨”琴，老杉木面板，柏木底板，通体黑金相间，乌木配件。琴腹内墨书“乙未秋昭闻斫于广陵，茗堂阿愚署”，琴底刻名“蕉林听雨”，下有韩春涛先生制“昭闻制琴”印一枚。龙池两侧对联为“一榻梦生琴上月，百花香入案头诗”。琴腰部有韩春涛制“美人君子共长生”印一枚。

"Listening to Rain in the Banana Forest" Qin made by ZhaoWen, Banana Leaf Style.

Overall length 120.0 cm, effective chord length 109.5 cm, forehead width 18.0 cm, shoulder width 19.5 cm, tail width 14.5 cm, and thickness 5.0 cm.

The banana leaf-style "Listening to Rain in the Banana Forest" Qin, whose surface panel is made of Chinese fir, and bottom panel cypress, has black and gold topcoat and ebony accessories. Inside the qin belly, it writes in ink "Made by ZhaoWen in the autumn of 2015 in Guangling, subscribed by A Yu of Mingtang". The bottom is inscribed with the name "Listening to Rain in the Banana Forest" and a seal that reads "Qin Made by ZhaoWen" made by Mr. Han Chuntao below. The lines "A couch dream born the qin and moon, fragrance of flowers fly into the poem." are on both sides of the Longchi. At the waist of the qin, there is the seal "Long Live Beauty and Gentleman" made by Han Chuntao.

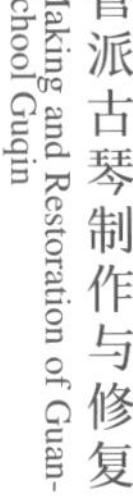

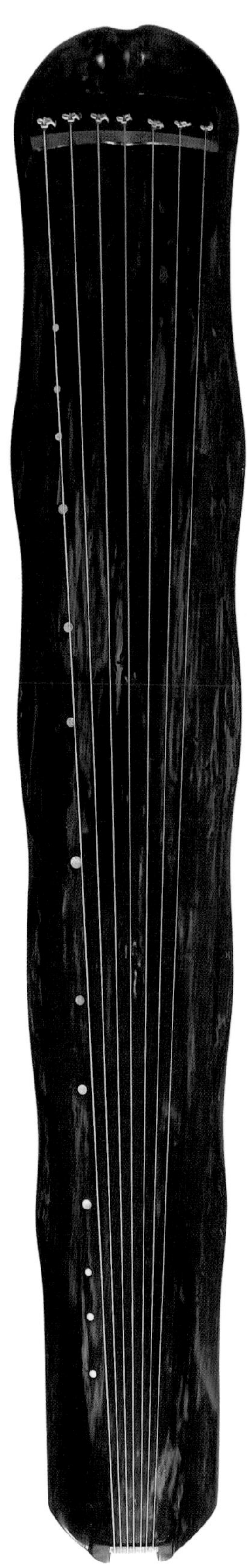

蕉林
聽雨
一榻夢生琴上月
百花香入案頭詩

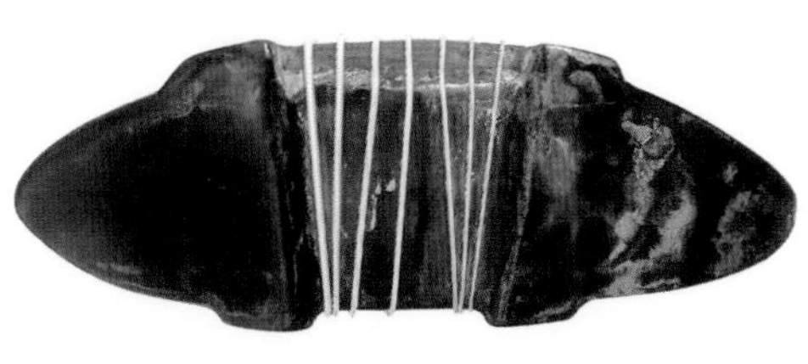

蕉林
聽雨

一榻夢生琴上月
百花香入案頭詩

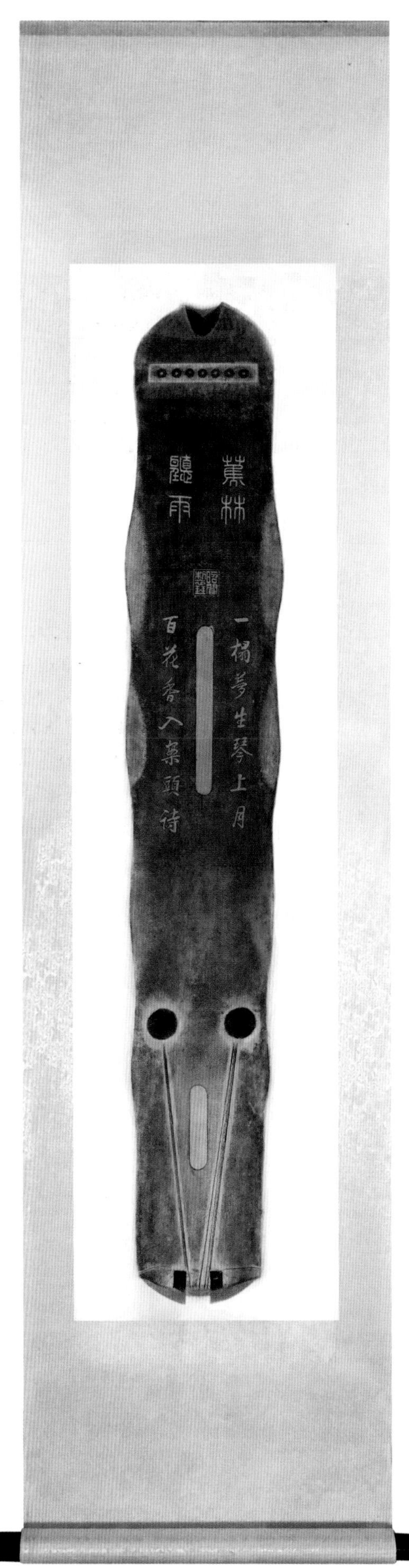
蕉林聽雨
一榻夢生琴上月
百花香入案頭詩

昭闻斫“昆山玉”琴 正合式

通长 121.6 厘米，有效弦长 111.2 厘米，额宽 20.5 厘米，肩宽 19.0 厘米，尾宽 14.7 厘米，厚 6.0 厘米。

正合式“昆山玉”琴，西安都城隍庙 660 年老桦木房梁面板，香椿木底板，通体黑漆透朱砂红，黑檀配件，龙池内可见古木沟壑旧痕。琴腹内墨书“乙未秋昭闻斫于长安，茗堂阿愚署”，琴底刻名“昆山玉”，下有韩春涛先生制“昭闻制琴”印一枚。龙池两侧对联为“赋就鹤来探句，琴清鱼出聆音”。琴腰部有“停云”印章一枚。

"Jade of Kunshan" Qin made by ZhaoWen, Zhenghe Style.

Overall length 121.6 cm, effective chord length 111.2 cm, forehead width 20.5 cm, shoulder width 19.0 cm, tail width 14.7 cm, and thickness 6.0 cm.

The Zhenghe-style "Jade of Kunshan" Qin's surface panel is made of a 660-year-old birch beam from the Capital City God Temple, and its bottom panel is made of Toona sinensis wood. The finishing coat is mainly black, blended with vermilion. It has ebony accessories, and old traces of ancient wood can be seen in its Longchi part. Inside the qin belly, it writes in ink "Made by ZhaoWen in the autumn of 2015 in Chang'an, subscribed by A Yu of Mingtang". The bottom is inscribed with the name "Jade of Kunshan" and a seal that reads "Qin Made by ZhaoWen" made by Mr. Han Chuntao below. The lines "When a poem is composed, a crane will fly all the way to hear; when the qin is played, fishes will jump out of the water to listen" are on both sides of the Longchi. A seal reading "Stop the Cloud" sits at the waist of the qin.

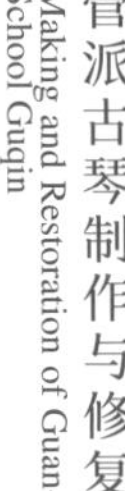

昆山玉
賦就鶴來探句
琴清魚出聆音

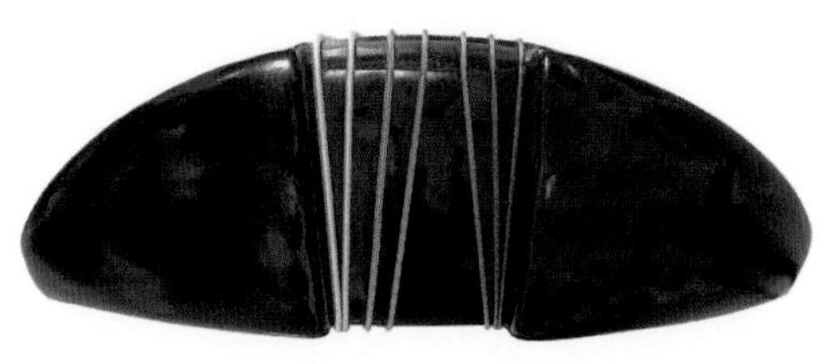

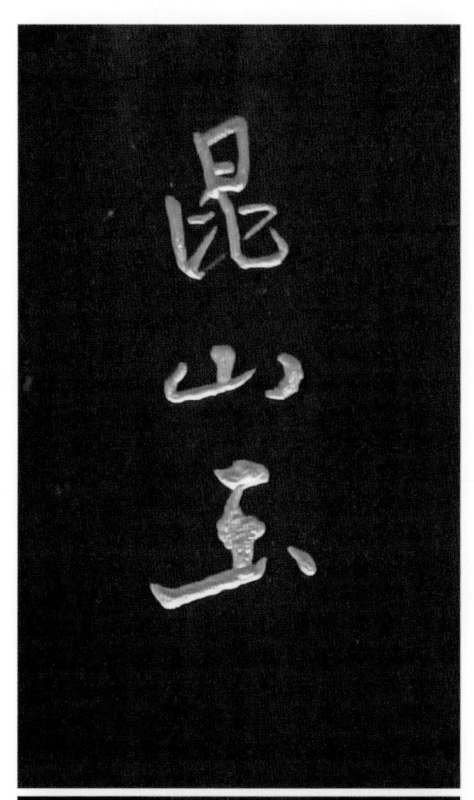
昆山玉

賦就鶴來採句
琴清魚出聆音

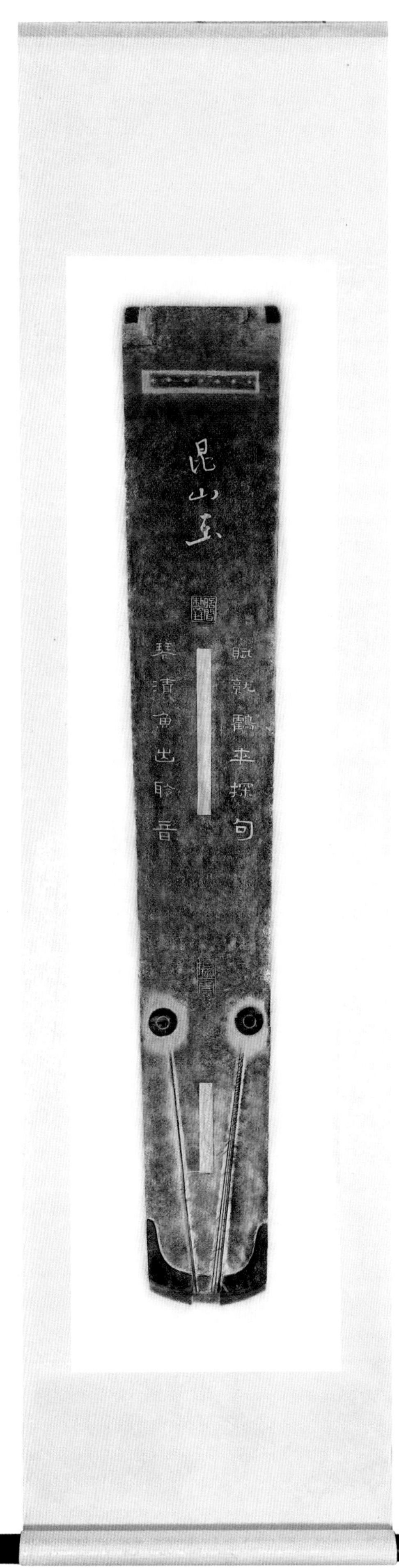
昆山玉

昭闻斫“清云”琴 宣和式

通长 123.5 厘米，有效弦长 111.5 厘米，额宽 19.5 厘米，肩宽 20.0 厘米，尾宽 14.6 厘米，厚 5.8 厘米。

宣和式“清云”琴，西安都城隍庙 660 年老松木房梁面板，香椿木底板，通体朱砂红，黑檀配件，龙池内可见古木沟壑旧痕。琴腹内墨书“丁酉春昭闻斫于长安，茗堂阿愚署”，琴底刻名“清云”，左侧落款“阿愚”，下有韩春涛先生制“昭闻制琴”印一枚。龙池两侧对联“抱琴看鹤去，枕石待云归”，由西安书画名家阿愚题写。琴腰部有“观众妙”印章一枚。

"Thin Clouds" Qin made by ZhaoWen, Xuanhe Style.

Overall length 123.5 cm, effective chord length 111.5 cm, forehead width 19.5 cm, shoulder width 20.0 cm, tail width 14.6 cm, and thickness 5.8 cm.

The Xuanhe-style "Thin Clouds" Qin's surface panel is made of a 660-year-old pine beam from the Capital City God Temple, and its bottom panel is made of Toona sinensis wood. The finishing coat is mainly red. It has ebony accessories, and old traces of ancient wood can be seen in its Longchi part. Inside the qin belly, it writes in ink "Made by ZhaoWen in the spring of 2017 in Chang'an, subscribed by A Yu of Mingtang". The name "Thin Clouds" is inscribed on the bottom, and on its left, there is a small seal reading "A Yu", together with a square seal made by Mr. Han Chuntao, which reads "Qin Made by ZhaoWen" below. The lines "Hold a qin to watch the crane, rest on a stone to wait for the cloud to return" on both sides of the Longchi are inscribed by A Yu, a master of calligraphy and painting in Xi'an. A seal reading "See the World" sits at the waist of the qin.

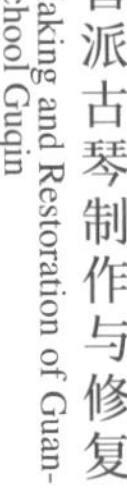

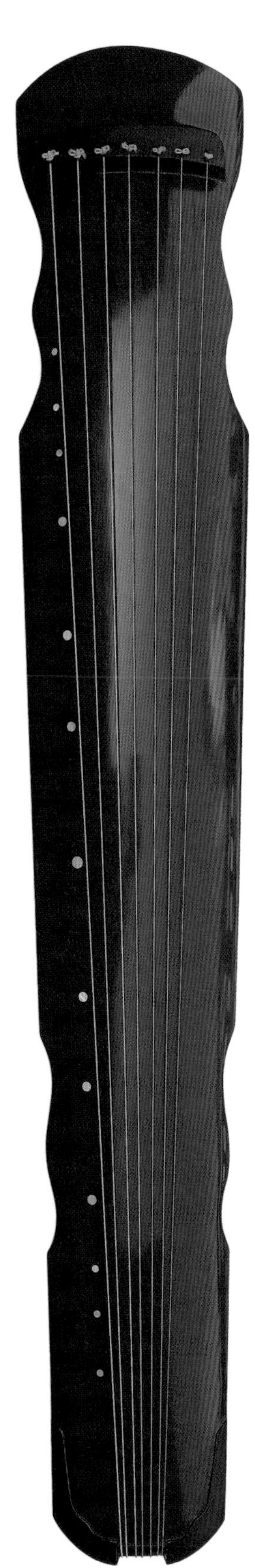

涛云
抱琴看鹤去
枕石待云归

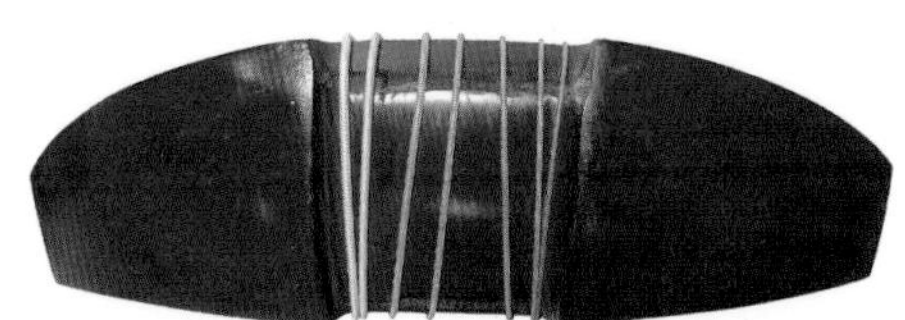

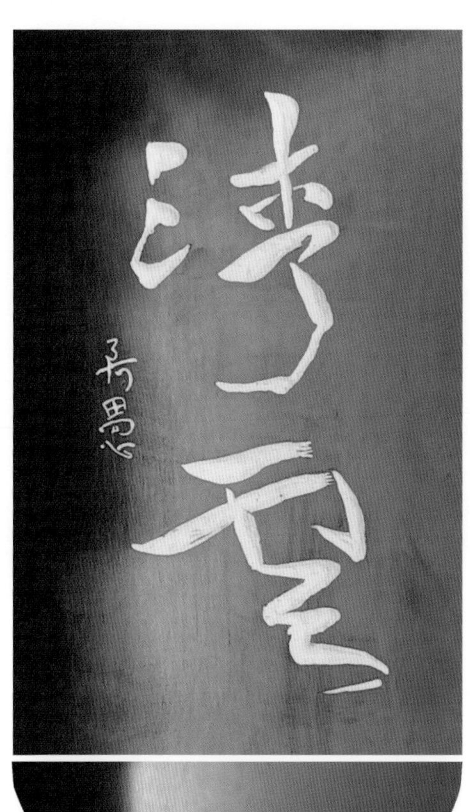

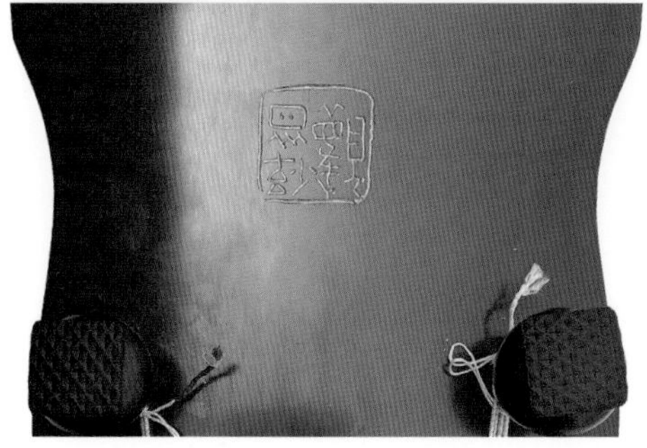

抱琴看鹤去
枕石待云归

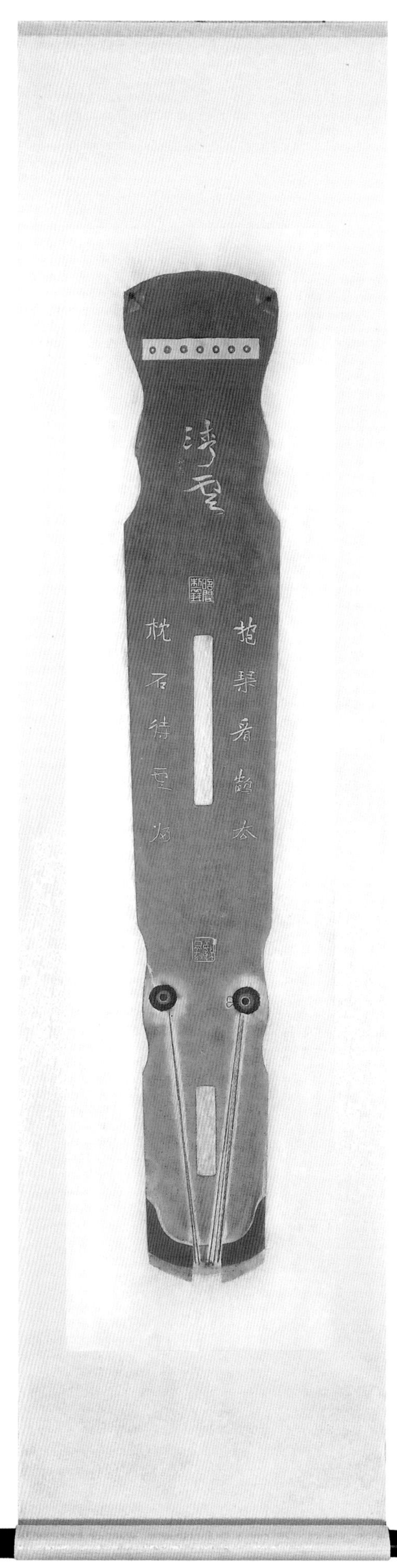
抱琴看鹤去
枕石待云归

昭闻斫"如影"琴 伏羲式

通长 96.6 厘米，有效弦长 86.8 厘米，额宽 16.5 厘米，肩宽 17.0 厘米，尾宽 12.8 厘米，厚 4.8 厘米。

伏羲式"如影"膝琴，杉木面板，西安都城隍庙 660 年老松木房梁底板，通体黑漆，黑檀配件。琴腹内墨书"丁酉荷月昭闻斫于终南山下百琴堂，茗堂阿愚署"，琴底刻名"如影"，下有韩春涛先生制"昭闻制琴"印一枚。

"Like Shadows" Qin made by ZhaoWen, Fuxi Style.

Overall length 96.6 cm, effective chord length 86.8 cm, forehead width 16.5 cm, shoulder width 17.0 cm, tail width 12.8 cm, and thickness 4.8 cm.

The Fuxi-style "Like Shadows" Qin surface panel is made of Chinese fir, and its bottom panel is made of a 660-year-old pine beam from the Capital City God Temple. The whole body is covered by black topcoat and ebony accessories. Inside the qin belly, it writes in ink "Made by ZhaoWen in the summer of 2017 in Baiqin Hall under Zhongnan Mountains, subscribed by A Yu of Mingtang". The name "Like Shadows" is inscribed on the bottom. Under the inscription, there is a seal made by Mr. Han Chuntao, which reads "Qin Made by ZhaoWen".

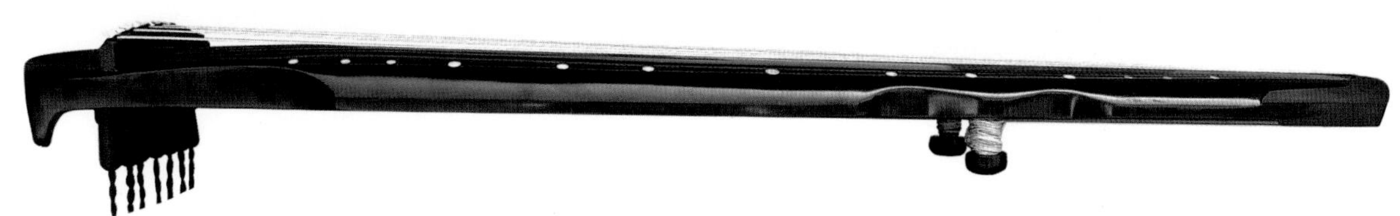

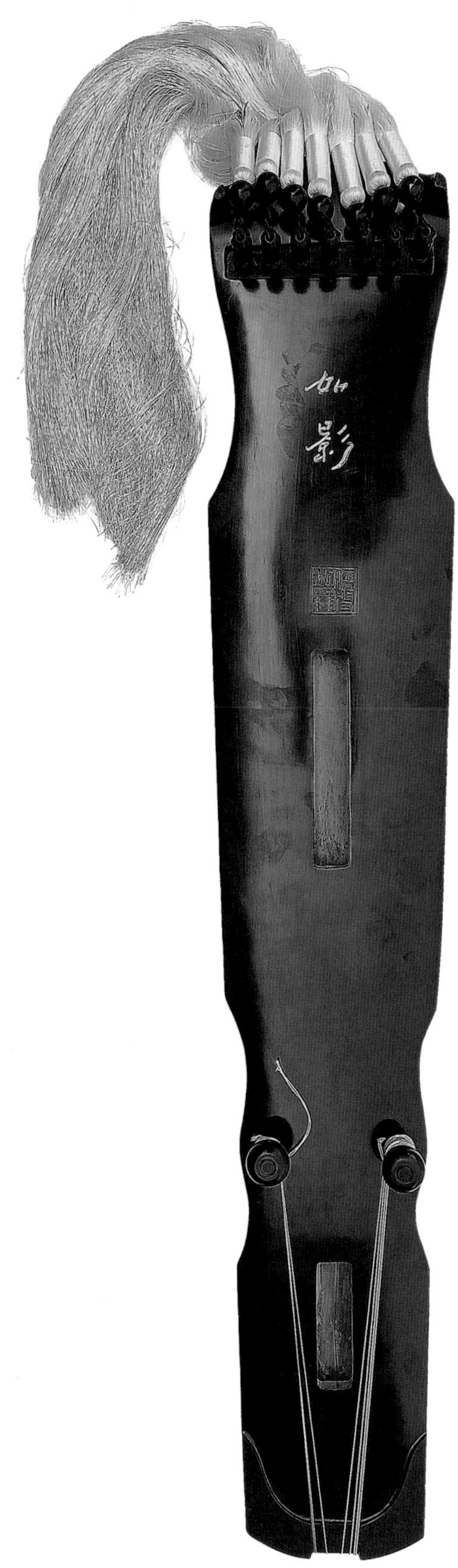
如影

如影

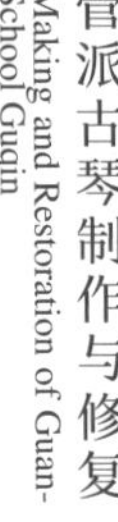

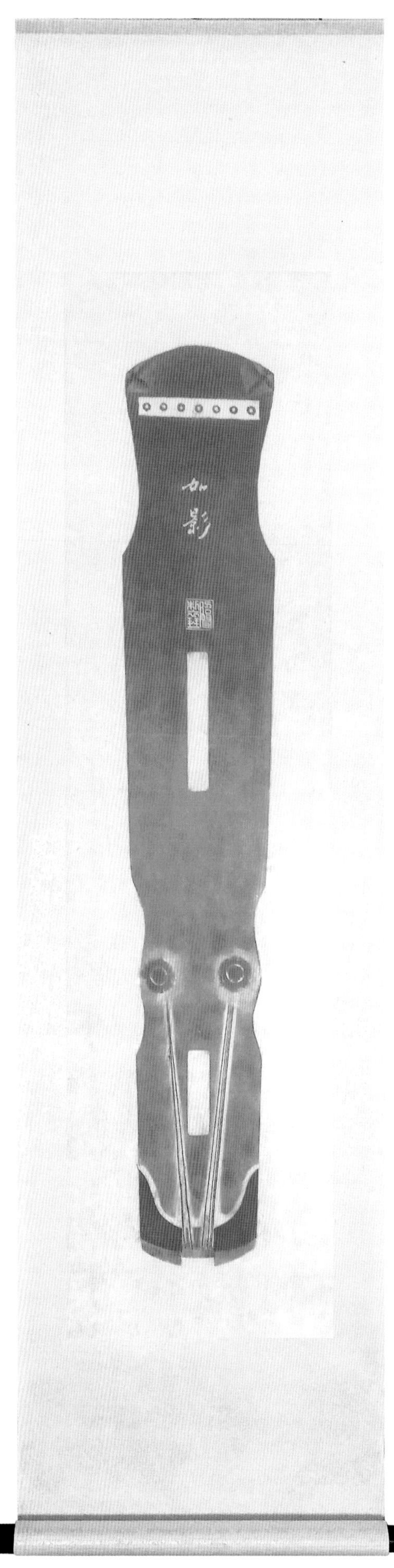
如影

后　记

古琴之雅在中国人的基因里生而有之，一张琴放在那里，即使你什么都不懂，也总有一种抑制不住的想去抚摸弹拨一下的冲动。没人知道周文王、周武王抚琴是为了教化于民还是为了抒发情感，总之他们为古琴文化奠定了雅、正的历史属性。自此，“文武七弦琴”在第一个被称之为“京”的地方诞生。三生有幸，我在这个被后来人称为“古琴发源地”的西安开办了中国古琴博物馆，也有了一片属于自己斫琴、写书的小天地。

认识田双坤恩师是在 2014 年的一个夏天，我与好友、在京琴家李新奇同游南锣鼓巷时，无意间闯入了他工作的小院儿。虽然我们之间的年龄相差近半百，但是一杯茉莉花茶和几句琴人间的家常，顿时拉近了彼此之间的距离。恩师作为管平湖先生唯一的斫琴弟子，尊重传统，低调务实，六十年来在北京那拥挤的小杂院儿里默默坚守着，从古琴的沉寂，到如今的兴盛，几经沉浮。如果跟他斫琴的日子久了，你就会发现他的一切工具、一切工序和一切数据，包括他讲出来的那些话语，哪一样没有管平湖先生的影子？这让你不禁感觉如同回到了 20 世纪 50 年代，恍若哪一天的傍晚，管平湖先生就会来到这个小院儿里，轻抚长褂，缓缓坐下，端起茶碗，手摇折扇，讲起那些工序与数据的道理所在。

2018 年的冬天，那时候我已经是陕西省古法斫琴非物质文化遗产代表性传承人，在中国艺术研究院音乐研究所董建国先生的见证下，我给田双坤恩师正式执拜师礼。他才是离管平湖先生最近的人，我有义务尽我所能记录下这一切，留给后世，以慰先贤。

特别说明：本书在采编期间，由于恩师年事已高，精力所限。故书中部分斫琴非核心工序，由本人与师兄弟姐妹在恩师的指导下完成，或进行示范拍摄。

2021 年 3 月

于中国古琴博物馆百琴堂

昭朋

Epilogue

When you have a Guqin in front of you, even if know nothing about it, there is always an uncontrollable impulse to touch and play it. This is the strong passion for this elegant art form that the Chinese nation is born with. No one knows exactly whether King Wen and Wu of Zhou played the qin to educate their people or just to express themselves. Yet they laid an elegant and positive historical attribute for the ancient qin culture. Since then, the "Seven-string Qin of King Wen and Wu" has been born in the first place called "Jing". I am incredibly lucky to set up the Chinese Guqin Museum in Chang'an, which is billed as the "birthplace of Guqin" by later generations, creating a small world for myself to make qins and write books.

I met my teacher Mr. Tian Shuangkun in the summer of 2014 when I was traveling with my friend Li Xinqi in the South Luogu Lane and accidentally entering his work studio. Despite our vast age difference of nearly half a century, a cup of jasmine tea and a few words immediately narrowed the distance between us. As Mr. Guan Pinghu's only disciple of Guqin making, Mr. Tian, who has been honored with the tradition, kept a low profile and stayed in his small, crowded courtyard in Beijing for 60 years, as Guqin shifted from silence to its heyday. Learning from him for a long while, I found that Mr. Guan Pinghu's shadow can find its expression in Mr. Tian in several aspects, whether the tools he used, or his processes and data in qin making, and even his words. That often brought me back to the days in the 1950s. It seemed that Mr. Guan Pinghu may come to this small courtyard in one evening in a long jacket, sit down slowly, pick up a tea bowl and a folding fan, and talk about those processes and data for qin making.

In the winter of 2018, after I became the representative inheritor of the intangible cultural heritage of the Guqin making in Shaanxi Province, under the witness of Mr. Dong Jianguo from the Music Research Institute of the Chinese Academy of Art, I held a ceremony to honor Mr. Tian Shuangkun as my teacher. No one is closer to Mr. Guan Pinghu than my teacher. So I take it my obligation to record all this story and leave it to future generations while comforting the sages.

Special note:

During the compilation of this book, due to the old age of my teacher whose energy was limited. Therefore, some of the non-core processes of Guqin production in this book were done by myself and my siblings under the guidance of my teacher or were photographed for demonstration purposes.

ZhaoWen
March 2021, at Baiqin Hall, Chinese Guqin Museum